FORD
FIESTA
1976 to February 1989

All Ford Fiesta models except Diesel engine versions.
Does not cover revised Fiesta range introduced in February 1989

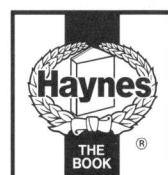

A Haynes Handbook
and Drivers Guide
© Haynes Publishing Group 1992

Printed and published by
J H Haynes & Co Ltd Sparkford
Nr Yeovil Somerset BA22 7JJ
England

ISBN 1 85010 816 1

British Library Cataloguing in Publication Data. A catalogue record of this book is available from the British Library

All rights reserved. No part of this book may be reproduced or transmitted in any form or by any means, electronic or mechanical, including photocopying, recording or by any information storage or retrieval system, without permission in writing from the copyright holder

HANDBOOK & DRIVERS GUIDE
by Steve Rendle

Full details of all servicing and repair tasks for the models covered by this Handbook can be found in the relevant Owners Workshop Manual – OWM **334** Ford Fiesta (1976 to 1983), or OWM **1030** Ford Fiesta (August 1983 to February 1989).

ACKNOWLEDGEMENTS

We gratefully acknowledge the assistance of RoSPA in compiling the information used in this Handbook. Thanks are also due to Duckhams Oils who provided lubrication data and to Britax who supplied the information on child safety. Certain illustrations, including the cover photography, are the copyright of the Ford Motor Company Limited and are used with their permission. Additional photographs supplied by Quadrant Picture Library/Auto Express. Thanks are also due to all those people at Sparkford who helped in the production of this Handbook.

We take great pride in the accuracy of information given in this Handbook, but vehicle manufacturers make alterations and design changes during the production run of a particular vehicle of which they do not inform us. No liability can be accepted by the authors or publishers for loss, damage or injury caused by any errors in, or omissions from, the information given.

FORD FIESTA

CONTENTS

About this Handbook	5
The Fiesta family	7
Buying and selling	9
General tips on buying	9
Points to look for when buying	10
General tips on selling	11
Dimensions and weights	13
Controls and equipment	15
Driver's instruments and controls	19
Interior equipment	26
Exterior equipment	28
Accidents and emergencies	29
How to cope with an accident	29
First aid	30
Requirements of the law	32
Essential details to record	32
Accident report form	33
How to cope with a fire	36
How to cope with a broken windscreen	36
What to do if your car is broken into	36
Breakdowns	37
Breakdowns on an ordinary road	37
Motorway breakdowns	38
Changing a wheel	38
Towing	40
Starting a car with a flat battery	41
What to carry in case of a breakdown	42
Driving safety	43
Before starting a journey	43
Driving in bad weather	44
Motorway driving	45
Towing a trailer or caravan	46
Alcohol and driving	48
Skid control	48
Advice to women drivers	49
Child safety	51
Driving abroad	53
Reducing the cost of motoring	57
Car crime prevention	59
Service specifications	61
Regular checks	65
Checking oil level	66
Checking coolant level	67
Checking brake fluid level	68
Checking tyres	68
Checking washer fluid level	70
Checking battery electrolyte level	70
Checking wipers and washers	71
Checking lights and horn	72
Checking for fluid leaks	72
Servicing	73
Service schedule	74
Safety first!	78
Buying spare parts	82
Service tasks	83
Seasonal servicing	101
Tools	104
What to buy	104
Care of your tools	104
Bodywork and interior care	105
Cleaning the interior	105
Cleaning the bodywork	105
Dealing with scratches	106
Bulb, fuse and relay renewal	107
Bulbs – exterior lights	108
Bulbs – interior lights	114
Fuses	114
Relays	116
Preparing for the MOT test	119
Fault finding	121
Car jargon	127
Local radio frequencies	137
Conversion factors	142
Distance tables	144
Index	146

FORD FIESTA

4 ABOUT THIS HANDBOOK

FORD FIESTA

ABOUT THIS HANDBOOK

The idea behind this Handbook is to help you to get the most out of your motoring.

Apart from the things that every owner needs to know, to deal with unexpected mishaps like a puncture or a blown light bulb, you'll find clearly-presented information on road safety, hints on driving abroad, and tips on how to prepare your car for the MOT test. We've also included details of local radio frequencies to help you avoid those inevitable 'jams', and there's a section on what to do in the unfortunate event of an accident.

For those not familiar with their car, there's a Section to explain the location and the operation of the various controls and instruments. Additionally, there's a Section on fault finding, and a useful glossary of car jargon.

Garage labour charges usually form the major part of any car servicing bill, and we hope to help you to reduce those bills by carrying out the more straightforward routine servicing jobs yourself. If you're about to start carrying out your own servicing for the first time, we aim to provide you with easy-to-follow instructions, enabling you to carry out the simpler tasks which perhaps you've left to a garage or a 'car-minded' friend in the past. Even if you prefer to leave regular servicing to a suitably qualified expert, by using this book you'll be able to carry out regular checks on your car to make sure that your motoring is safe and hopefully trouble-free. You'll also find advice on buying suitable tools, and safety in the home workshop.

Some readers of this Handbook may not yet have bought a Fiesta, so we've included a brief history of the range, and some useful tips on buying and selling.

All in all, we hope that this book will prove a handy companion for your motoring adventures, and hopefully we'll help to reduce the problems which inevitably crop up in everyday driving. If you're bitten by the DIY bug, and you're keen to tackle some of the more advanced repair jobs on your car, then you'll need our **Owners Workshop Manual** for your particular model (OWM 334 for Fiesta 'Mark 1' models, or OWM 1030 for Fiesta 'Mark 2' models). These manuals give a step-by-step guide to all the repair and overhaul tasks, with plenty of illustrations to make things even clearer.

Happy motoring!

THE FIESTA FAMILY

▲ *Fiesta 950 Base (1976)*

▲ *Fiesta 1.1 L (1984)*

THE FIESTA FAMILY

The Fiesta was first introduced into the UK in February 1976, and was especially significant, as it was the first front-wheel-drive production car to be marketed by Ford in the UK.

All models in the Fiesta range are of 3-door Hatchback design.

To begin with, only 950 cc & 1.1 litre engines were available, but a 1.3 litre engine was soon introduced to provide more power, and in 1981 the high performance XR2 model was introduced, powered by a 1.6 litre engine.

In August 1983, the whole range was facelifted and revised to incorporate many improvements over the original design, and these later Fiesta models are often referred to as 'Mark 2's. The major changes introduced with the 'Mark 2' models were the smoother, more aerodynamic body shape, and the 1.3 & 1.6 litre CVH type engines, which replaced the older OHV type 1.1 & 1.3 litre engines used in the 'Mark 1' models. In 1986, a 1.4 litre CVH type engine superseded the 1.3 litre CVH type.

Models were produced with 4- and 5-speed manual gearboxes and CTX automatic transmission, and with a number of different trim levels, from the Base 950 cc model, to the sporting XR2 model.

All members of the Fiesta family share a similar mechanical layout, and they all share the same family of engines and gearboxes, although many detailed modifications and improvements were made during the 13 years of production.

Most of the mechanical components of the vehicles are fairly conventional, and the main systems are designed to keep servicing time and costs as low as possible. The clutch, brakes, wheel bearings, etc are all routine-maintenance-free, and should only require attention when certain items need to be renewed.

All the engines are of 4-cylinder in-line layout, and are mounted transversely at the front of the car, with the gearbox positioned on the left-hand end of the engine (when viewed from the driver's seat).

With the wide availability of spare parts, these cars are ideal for the enthusiastic amateur mechanic, who wants to keep running costs to a minimum.

▲ *Fiesta 1.3 Ghia (1984)*

8 THE FIESTA FAMILY

▲ *Fiesta XR2 (1981)*

▲ *Fiesta XR2 (1985)*

BUYING AND SELLING

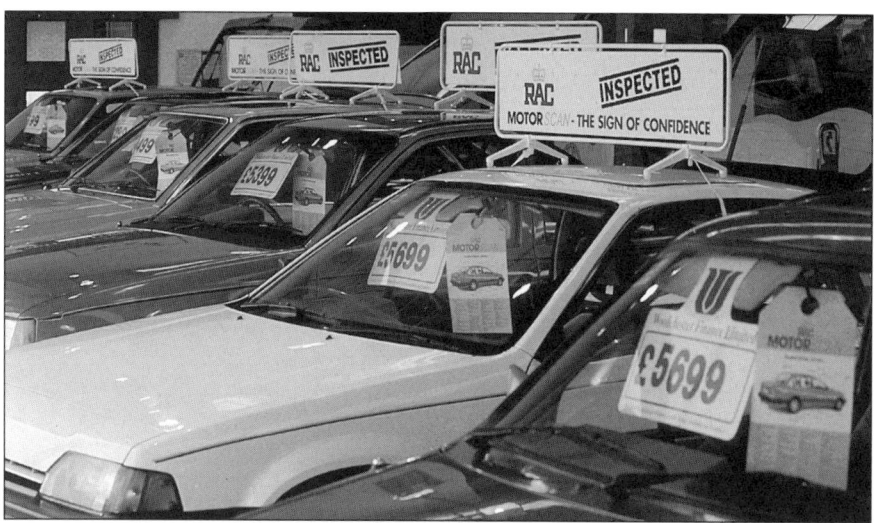

GENERAL TIPS ON BUYING
- Don't rush out and buy the first car to catch your attention
- Always buy from a recognised dealer in preference to buying privately
- Have the car checked over by someone knowledgeable before you buy it
- In general, you get what you pay for!

Before buying a second-hand car, it's worthwhile doing some homework to try and avoid some of the pitfalls waiting for the unwary. First of all, don't rush out and buy the first car to catch your attention (all that glitters is not gold!), and remember that much of the responsibility is yours when it comes to the soundness of the deal, especially when buying a car privately.

Wherever possible, buy from a recognised dealer, and check that the dealer is a member of the Retail Motor Industry Federation, as this will provide you with certain legal safeguards if you have any problems. If you buy from a dealer, you are covered by the Sale Of Goods Act, which in summary states that the goods must be fit for their intended purpose, the goods must be of proper quality, and the goods must be as described by the seller. If you're buying a car privately, ask to see the service receipts and the MOT certificates going back as long as the car has been in the possession of the current owner (this will help to establish that the car hasn't been stolen, and that the recorded mileage is genuine), and always ask to view the car at the seller's private address (to make sure that the car isn't being sold by an unscrupulous dealer posing as a private seller). Check the vehicle documents for obvious signs of forgery and, if in doubt, contact the DVLA and give them details of the Registration Document, as they will be able to run a check on its authenticity.

As far as the soundness of the car itself is concerned, a genuine service history is helpful. This is provided by the service book supplied with the car when new, which should be completed and officially stamped by an authorised garage after each service. Cars with a full service history (fsh) usually command a higher price than those without.

To check the condition of a car, a professional examination is well worthwhile, if you can afford it, and organisations such as the AA and RAC will be able to provide such a service. Otherwise, you must trust your own judgement, and/or that of a

BUYING AND SELLING

service. Otherwise, you must trust your own judgement, and/or that of a knowledgeable friend. Although it's tempting, try not to overlook the mechanical soundness of the car in favour of the overall appearance. It's relatively easy to clean and polish a car every week, but when were the brakes and tyres last checked? Above all, safety must always take priority. Don't view a car in the wet, as water on the bodywork can give a misleading impression of the condition of the paintwork. First of all, check around the outside of the car for rust, and for obvious signs of new or mismatched paintwork which might show that the car has been involved in an accident. Check the tyres for signs of unusual wear or damage, and check that the car 'sits' evenly on its suspension, with all four corners at a similar height. Open the bonnet and check for any obvious signs of fluid leakage (oil, water, brake fluid), then start the engine and listen for any unusual noises – some background noise is to be expected on older cars, but there should be no sinister rattles or bangs! Also listen to the exhaust to make sure that it isn't 'blowing' indicating the need for renewal, and check for signs of excessive exhaust smoke. Black smoke may be caused by poorly adjusted fuel mixture, which can usually be rectified fairly easily, but blue smoke usually indicates worn engine components, which may prove expensive to repair. Finally, drive the car, and test the brakes, steering and gearbox. Make sure that the car doesn't pull to one side, check that the steering feels positive, and ensure that the gears can be selected satisfactorily without undue harshness or noise. Listen for any unusual noises or vibration, and keep an eye on the instruments and warning lights to make sure that they are working and indicating correctly.

If all proves satisfactory, try to negotiate a suitable deal, but remember that a dealer has to work to a profit margin, and it's unlikely that you'll find a good car for a silly price. Always obtain a receipt for your money.

On the whole, it's true to say that you get what you pay for. Above all, don't be rushed into making a hasty decision.

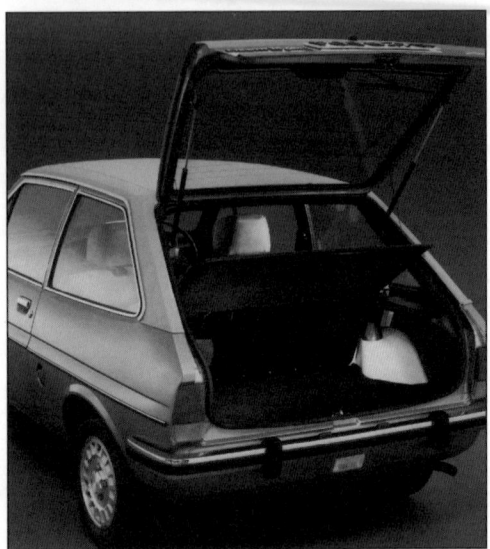

POINTS TO LOOK FOR WHEN BUYING A FIESTA

Generally, the Fiesta is a reliable car, and is easy to drive and maintain. Performance is average for this class of car, although Fiestas tend to be less economical than many of their rivals. The 5-speed gearbox introduced with the 'Mark 2' models can give a noticeable improvement in fuel economy compared with the 4-speed gearbox.

In general, the later 'Mark 2' models built from August 1983 are an all-round improvement on the earlier models, and will prove a better buy if your budget allows.

Corrosion can be a problem on early models, and there are one or two particular problem areas which should be checked. Generally, the later 'Mark 2' models are less prone to rust than their earlier 'Mark 1' counterparts.

The biggest problem area for rust is underneath the car. Look under the floor, just in front of each rear wheel arch, where the floor panel joins the panel which curves up under the rear seats. If any serious rust is visible, it's likely that the car will fail the MOT test. Although the damage can be repaired by welding, repairs can be expensive, and it's best to reject the car if rust is apparent.

Open the bonnet and check the inside

BUYING AND SELLING

surfaces of the front wings either side of the engine compartment. At the same time, check the area around the battery, which is prone to corrosion due to battery acid being washed over the metal by rain water.

Other potential problem areas are the front and rear body panels behind the bumpers, the sill panels below the front doors, and the bottoms of the doors themselves. The headlight reflectors are also prone to corrosion, and this can be expensive to repair, as the only way to solve the problem is to fit a new headlight.

It should not be difficult to find a rust-free car, and if you find more than two problem areas, it's best to reject the car.

The engines fitted to Fiestas are generally reliable, and may be of either OHV or CVH type (refer to 'Servicing' on page 83 for details of how to identify engines).

On models with OHV type engines, open the bonnet and check for obvious signs of oil or coolant leakage, particularly around the top of the engine (through the cylinder head gasket). Start the engine and listen for a pronounced rattling noise. Some noise is to be expected on high-mileage engines, but if the noise is excessive, the engine will almost certainly require attention, and the car should be rejected.

CVH type engines are prone to problems with a build-up of black sludge inside the engine if regular oil changes are not carried out. To check for this, remove the oil filler cap from the top of the engine and run a finger around the inside of the hole. If sludge is present, reject the car. Another point to note for CVH engines is that the camshaft drivebelt should be renewed after 36 000 miles. If you're not sure whether this has been done, it's sensible to have the necessary work carried out as soon as you buy the car, as if the drivebelt breaks in service, very serious and expensive engine damage can occur.

On XR2 models, the front brake and suspension components are prone to wear. To check for this, apply the brakes firmly when driving at speed, and if the car vibrates, particularly through the steering wheel, reject the car.

The Ford carburettor fitted to some models can cause starting problems and poor fuel economy, and it's common practice to replace this type of carburettor with a Weber type. This is a worthwhile modification, and is relatively inexpensive.

Other possible problem areas are the contact breaker type distributor on early models which is prone to failure, and the self-adjusting clutch mechanism (fitted to all models) which can cause problems when the components wear.

Fiestas are plentiful on the secondhand market, and it shouldn't be difficult to find a good car which will give trouble-free service.

GENERAL TIPS ON SELLING

- Make sure that the car is clean and tidy
- Make sure that all of the service documents, registration document etc, are available for inspection
- Ask yourself ... 'Would I buy this car?'

Obviously when selling a car, bear in mind the points which the prospective buyer should be looking for, as described in the above sections. It goes without saying that the car should be clean and tidy, as first impressions are important. Any fluid leaks should be cured, and there's no point in trying to disguise any major bodywork or mechanical problems.

If you're trading the car in with a dealer, you will always get a lower price than if you sell privately, but you can be fairly sure that there will be less comeback to you should any unexpected problems develop. If selling privately, don't allow the buyer to take the car away until you have his/her money, and it's a good idea to ask him/her to sign a piece of paper to say that he/she is happy to buy the car as viewed, just in case any problems develop later on. Give a receipt for the money paid.

FORD FIESTA

DIMENSIONS AND WEIGHTS

FORD FIESTA

DIMENSIONS AND WEIGHTS

Note: All figures are approximate, and will vary depending on model

DIMENSIONS [mm (in)]
Overall length

'Mark 1' models (without bumper overriders)	**3565** (140.5)
'Mark 1' models (with bumper overriders)	**3609** (142.2)
'Mark 2' models except XR2 (without bumper overriders)	**3648** (143.7)
'Mark 2' models except XR2 (with bumper overriders)	**3695** (145.6)
'Mark 2' XR2 models	**3712** (146.3)

Overall width

'Mark 1' models	**1567** (61.7)
'Mark 2' models except XR2	**1585** (62.4)
'Mark 2' XR2 models	**1620** (63.8)

Overall height

All models	**1314** to **1334** (51.8 to 52.6)

WEIGHTS [kg (lb)]
Nominal kerb weight

'Mark 1' models	**700** to **780** (1544 to 1720)
'Mark 2' models	**760** to **840** (1676 to 1852)

Maximum roof rack load

All models	**75** (165)

Maximum towing weight (for trailer or caravan)

All 'Mark 1' models except 950 cc models	**800** (1764)
950 cc 'Mark 1' models	**540** (1191)
'Mark 2' models with 4-speed manual gearbox	**540** (1191)
'Mark 2' models with 5-speed manual gearbox	**800** (1764)

Note: Towing is not recommended for models fitted with CTX automatic transmission

Maximum trailer/caravan noseweight

All models	**50** (110)

FORD FIESTA

14 CONTROLS AND EQUIPMENT

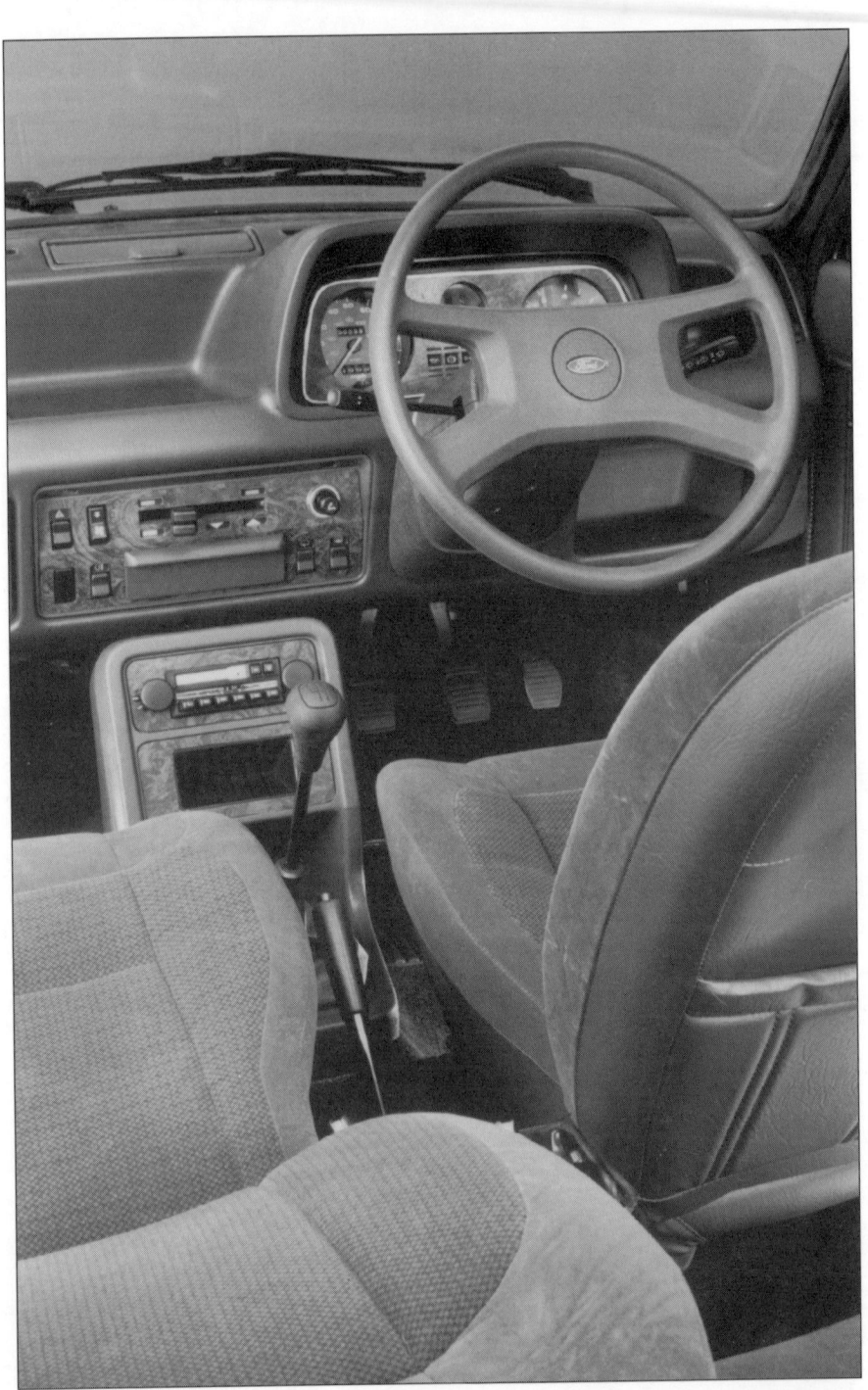

FORD FIESTA

CONTROLS AND EQUIPMENT

For those not familiar with the Fiesta models, this Section will help to identify the instruments and controls. Typical instrument panel layouts are shown in the accompanying illustrations. The operation of most equipment is self-explanatory, but some items require further explanation to ensure that their use is fully understood.
Note that not all items are fitted to all models.

▲ *Typical instrument panel layout for 'low series' Fiesta 'Mark 1' models*

1   Speedometer
2   Fuel gauge
3   Temperature gauge
4   Hazard flasher switch
5   Heating and ventilation controls
6   Tailgate wash/wipe switch
7   Heated rear window switch
8   Cigarette lighter
9   Ashtray
10 Rear foglight switch
11 Windscreen washer pump
12 Low oil pressure warning light
13 Handbrake 'on' warning light
14 Direction indicator warning light
15 Headlight main beam warning light
16 Ignition warning light
17 Bonnet release lever

FORD FIESTA

16 CONTROLS AND EQUIPMENT

▲ *Typical instrument panel layout for 'high series' Fiesta 'Mark 1' models*

1. Hazard flasher switch
2. Heating and ventilation controls
3. Cigarette lighter
4. Trip meter reset button
5. Speedometer
6. Low oil pressure warning light
7. Handbrake 'on' warning light
8. Temperature gauge
9. Fuel gauge
10. Headlight main beam warning light
11. Ignition warning light
12. Tachometer ('rev. counter')
13. Instrument illumination control
14. Rear foglight switch
15. Ashtray
16. Clock
17. Radio
18. Tailgate wash/wipe switch
19. Heated rear window switch
20. Choke control
21. Direction indicator warning light
22. Ignition switch/steering lock
23. Bonnet release lever

FORD FIESTA

CONTROLS AND EQUIPMENT   17

▲ *Typical instrument panel layout for 'low series' Fiesta 'Mark 2' models*

1 Digital clock
2 Headlight main beam warning light
3 Sidelight/tail light warning light
4 Temperature gauge
5 Fuel gauge
6 Glow plug warning light (Diesel models only)
7 Speedometer
8 Heated rear window switch
9 Tailgate wash/wipe switch
10 Rear foglight switch
11 Fusebox
12 Choke control
13 Ignition switch/steering lock
14 Ignition warning light
15 Hazard flasher switch
16 Direction indicator warning lights
17 Low oil pressure warning light
18 Low brake fluid level warning light
19 Ashtray
20 Cigarette lighter
21 Booster fan switch
22 Air temperature control
23 Air distribution control

FORD FIESTA

18 CONTROLS AND EQUIPMENT

▲ Typical instrument panel layout for 'high series' Fiesta 'Mark 2' models

1 Radio
2 Sidelight/tail light warning light
3 Headlight main beam warning light
4 Tachometer ('rev. counter')
5 Temperature gauge
6 Fuel gauge
7 Speedometer
8 Trip meter reset button
9 Heated rear window switch
10 Tailgate wash/wipe switch
11 Rear foglight switch
12 Fusebox
13 Choke control
14 Ignition switch/steering lock
15 Low oil pressure warning light
16 Direction indicator warning lights
17 Hazard flasher switch
18 Ignition warning light
19 Handbrake 'on' warning light
20 Low brake fluid level warning light
21 Loudspeaker balance control
22 Air temperature control
23 Booster fan switch
24 Air distribution control
25 Ashtray
26 Cigarette lighter

FORD FIESTA

CONTROLS AND EQUIPMENT 19

DRIVER'S INSTRUMENTS AND CONTROLS

Speedometer

Indicates the car's road speed, and incorporates a mileage recorder. Some models have a trip meter, which can be reset by pressing the button protruding from the instrument.

Tachometer ('rev. counter')

Indicates engine speed in revolutions per minute (rpm or rev/min). For normal driving and the best fuel economy, the engine speed should be kept in the 2000 to 4000 rpm range. The red band indicates the maximum permissible engine speed.

Fuel gauge

Indicates the quantity of fuel remaining in the tank. On 'Mark 1' models, when the needle enters the orange band, there are approximately 4.5 litres (1 gallon) of fuel remaining. On 'Mark 2' models, when the needle enters the red band, there are approximately 8 litres (1.8 gallons) of fuel remaining.

Temperature gauge

Indicates the engine coolant temperature. As the engine warms up, the needle should move from the blue (cold) end of the scale into the middle region. If the needle remains in the blue section, or enters the red (hot) section, a fault is indicated, and advice should be sought (refer to 'Fault finding' on page 123).

Do not continue to run an engine which shows signs of overheating

Clock

ANALOGUE TYPE

To adjust the time, push in the button and turn it.

DIGITAL TYPE

On models with a facia-mounted clock, the time is displayed in the 12-hour format, and can be viewed with the ignition switched off by pressing and holding the left-hand adjustment button.

On models with a roof-mounted clock, the time is displayed in the 24-hour format, and the display is illuminated with either the ignition or the lights switched on.

To adjust the time, switch on the ignition. Press the button on the left to change the hour setting, and press the button on the right to change the minutes setting. To synchronise the clock with an exact time signal, press and hold both buttons, then release the buttons at the time signal, and set the time as described previously. On some models, a pointed instrument such as a ballpoint pen may be required to press the adjustment buttons.

▲ Digital clock
h Hours adjustment button
m Minutes adjustment button

MULTI-FUNCTION DIGITAL TYPE (EARLY MODELS)

The clock can display the date, and has a timer feature. To operate the clock, the ignition must be switched on.

Adjusting time

To change the hours setting, press the upper recessed button **(F)** once, using a pointed instrument such as a ballpoint pen. The display will show the hour number on the left, and either 'A' for am or 'P' for pm on the right. The hour number will increase by one each time the button is pressed again. Four seconds after the button is last pressed, the display will show normal time, but with the newly-selected hour number.

To change the minutes setting, press the lower recessed button **(G)** once. The display will still show normal time, but the clock will have stopped, and the minute number will

CONTROLS AND EQUIPMENT

increase by one each time the button is pressed again. To restart the clock after setting the time, press the **stop/start** button **(C)**. If the clock is set slightly ahead of actual time, it can be synchronised with a time signal by pressing the **stop/start** button on the signal.

Date display

To display the date, press the **select** button **(A)** once. The date will be displayed for four seconds, after which the time will reappear.

Adjusting date

To set the date, obtain the date display as described previously. To set the day, immediately press the top recessed button **(F)** once. The date will increase by one each time the button **(F)** is pressed. To set the month, press the bottom recessed button **(G)** immediately after obtaining the date display. The month number will increase by one each time button **(G)** is pressed. Four seconds after either recessed button is released, the display will show normal time.

Timer

To operate the timer, with the display showing the time, press the **select** button **(A)** twice. The timer can now be started by pressing the **stop/start** button **(C)**. Pressing the **stop/start** button again will stop the timer and freeze the display. A third press will restart the timer from where it stopped. The timer will count to 59 minutes 59 seconds, and then roll over to 1 hour 00 minutes and continue to count at one minute intervals. When the timer reaches 59 hours 59 minutes, it automatically resets to 0 and continues counting.

Either of the two other displays can be selected when the timer is working. To display the time, press the **select** button once, and to display the date, press the **select** button again. To reset the timer to zero with the clock showing time, press the **select** button twice to display the timer, then to zero the timer press the **reset** button **(B)**. The display can now be set to show the normal time by pressing the **select** button, or the timer can be restarted by pressing the **stop/start** button. The recorded time, when the timer was stopped, will be stored in the memory until the timer is restarted again, even if the ignition is switched off. If the ignition is switched off when the timer is still running, it will continue to count until the ignition is switched on again and the timer is stopped or reset; the timer can therefore be used to record actual journey time.

▲ Multi-function digital clock
  A Select button
  B Reset button
  C Stop/start button
  D Timer symbol
  E Colon
  F Recessed adjustment button (early models only)
  G Recessed adjustment button (early models only)

MULTI-FUNCTION DIGITAL TYPE (LATER MODELS)

The clock can display the date, and has a timer feature. On models up to 1986, no display is shown with the ignition switched off, but the display can be recalled for four seconds by pressing the **select** button **(A)**. One press will display time, and further presses will display the date and timer. On models from 1986, the display is illuminated with either the ignition or the lights switched on, and the date and timer functions can be displayed by pressing the **select** button as described previously. The ignition must be switched on to carry out all setting and adjustments.

Adjusting time and date

The time and date must be set using the following sequence. With the display showing the time, press the **reset** button **(B)** once. The display should flash showing '12H' or '24H' depending on whether the display has been set to the 12- or 24-hour format. Press the **stop/start** button **(C)** to change the format. Press the **reset** button again. The hour number

FORD FIESTA

CONTROLS AND EQUIPMENT

on the left will flash. If the 24-hour format was selected, the minute number on the right will remain stationary, but if the 12-hour format was selected, an 'A' for am or 'P' for pm will appear in place of the minutes. Press the **stop/start** button to increase the hour number as required. Press the **reset** button again. The minute number on the right of the display will now flash. Press the **stop/start** button to increase the minute number. Press the **reset** button again. The date will now be displayed, with the day number flashing on the left, and the month number stationary on the right. Press the **stop/start** button to increase the day number. Press the **reset** button again. The display now shows the date with the day stationary on the left, and the month number flashing on the right. Press the **stop/start** button to increase the month number. When the adjustments are complete, press the **reset** button to return to the time display. To return to the time display at any point during the procedure, press the **select** button.

Date display

To display the date, press the **select** button **(A)** once. The date will be displayed for four seconds, after which the time will reappear.

Timer

The operation of the timer is the same as described for the earlier type of multi-function clock.

Warning lights

These lights warn the driver of a fault, or inform the driver that a particular device is in operation.

LOW BRAKE FLUID LEVEL WARNING LIGHT

Warns that the brake fluid level is low. If the light comes on while driving, stop immediately and check the brake fluid level (refer to *Regular checks* on page 68).

Do not continue to drive the car if a brake fluid leak is suspected.

HANDBRAKE 'ON' WARNING LIGHT

Warns that the handbrake is applied. Note that on some 'Mark 1' models, this light also acts as a low brake fluid level warning light.

LOW OIL PRESSURE WARNING LIGHT

Warns that the engine oil pressure is low. If the light stays on for more than a few seconds after start-up, it's likely that the engine is worn, and advice should be sought. If the light comes on whilst driving, switch off the engine immediately and seek advice.

IGNITION WARNING LIGHT

Acts as a reminder that the ignition circuit is switched on if the engine isn't running, and acts as a no-charge warning light. If the light stays on after starting, or comes on whilst driving, the battery is not charging properly. The battery may therefore become fully discharged (resulting in a 'flat' battery), and the best course of action is to stop and seek advice.

HEADLIGHT MAIN BEAM WARNING LIGHT

Acts as a reminder that the headlight main beam is switched on.

SIDELIGHT/TAIL LIGHT WARNING LIGHT

Acts as a reminder that the sidelights and tail lights are switched on.

DIRECTION INDICATOR WARNING LIGHT(S)

Show(s) that the direction indicators are switched on.

ECONOLIGHTS

These lights provide a rough indication as to how economically the car is being driven. As a guide, aim to drive so that the lights come on only momentarily, or not at all. The amber light will come on when the car is being driven less economically, and the red light will come on when the car is being driven uneconomically. In practice, it will be very difficult to drive without causing the lights to come on at some stage, especially when accelerating.

CONTROLS AND EQUIPMENT

Ignition switch/steering lock

The switch has four positions as follows:
- O Ignition off, steering locked
- I Ignition off, accessory circuits on, steering unlocked
- II Ignition on, and all electrical circuits on
- III Starter motor operates (release the key immediately the engine starts)

▲ Ignition switch/steering lock key positions

Choke

The choke control should be pulled out as necessary to start a cold engine, and should be gradually pushed in as the engine warms up, until the car can be driven smoothly with no choke. A warning light incorporated in the choke control knob indicates that the choke is in operation. Note that some models have an automatic choke.

Gearbox

MANUAL GEARBOX

Either a 4- or 5-speed gearbox may be fitted. The gear positions follow the usual 'H' pattern, but the method of selecting reverse gear varies depending on model:

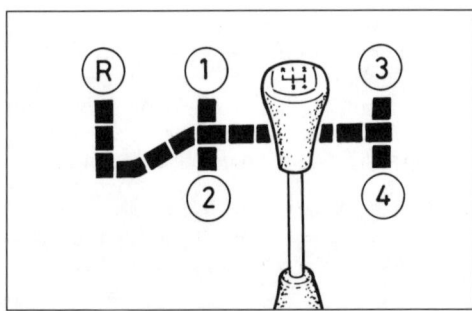

▲ Gear lever positions ('Mark 1' models)

'Mark 1' models

Reverse is positioned to the left of 1st gear. Reverse is selected by moving the lever fully left from the neutral position, then pressing down against spring pressure and pushing further left and forwards.

'Mark 2' models

4-speed gearbox – Reverse gear is positioned to the left of 1st gear. Reverse is selected by moving the lever fully left from the neutral position, then pressing down against spring pressure and pushing the lever forwards.

5-speed gearbox – Reverse gear is positioned opposite 5th gear. On models built before February 1987, reverse is selected by moving the lever fully right from the neutral position, then pushing down against spring pressure and pulling rearwards. On models built after February 1987, reverse is selected by moving

▲ Gear lever positions ('Mark 2' models)
- A 4-speed gearboxes
- B 5-speed gearboxes up to February 1987
- C 5-speed gearboxes from February 1987

CONTROLS AND EQUIPMENT

the lever fully right from the neutral position against spring pressure before pulling rearwards.

CTX AUTOMATIC TRANSMISSION

There are five selector positions:
  P – locks the transmission, and should be used when parking. The engine can be started. Only select **P** after parking when the car is stationary, and always apply the handbrake.
  R – selects **Reverse** gear.
  N – selects **Neutral**. The engine can be started. No power is transmitted to the wheels.
  D – selects **Drive**. The transmission will infinitely vary the gear ratio between the lowest and highest available.
  L – gives increased engine braking. Should be used when driving up or down steep slopes.

The handbrake or footbrake must be applied before selecting **R**, **D**, or **L** when the car is stationary. To select positions **R** and **P**, the selector lever must first be pushed down in the **N** position. The lever must also be pushed down to move it out of the **P** position. To move out of the other positions, the selector lever must be moved sideways first.

Multi-function switch

Controls the direction indicators, headlight flash & main beam, and horn. On models fitted with driving lights, the driving lights operate with the headlight main beam and headlight flash.

▲ *Multi-function switch*
  1 Horn
  2 Right direction indicator
  3 Left direction indicator
  4 Headlight main beam
  5 Headlight flash

Exterior lights switch

Controls the sidelights/tail lights and headlights.

▲ *Exterior lights switch*
  1 Sidelights/tail lights
  2 Headlights

Windscreen wash/wipe switch

Controls the windscreen wipers and washers. Where fitted, the headlight washers will operate at the same time as the windscreen washers, provided the headlights are switched on.

▲ *Windscreen wash/wipe switch*
  1 Normal wiper speed
  2 Fast wipe
  3 Single wipe or intermittent wipe (depending on model)
  4 Windscreen washers (not applicable to certain 'Mark 1' models)

Tailgate wash/wipe switch

Depress the switch to the first position to operate the wiper. Depress the switch further and hold it depressed to operate the washer (Note that some 'Mark 1' models have a foot-operated windscreen washer pump).

CONTROLS AND EQUIPMENT

Windscreen washer pump ('low series Mark 1' models)

On some 'Mark 1' models, the windscreen washer pump is foot-operated, and is fitted to the left of the clutch pedal.

Heating and ventilation controls ('Mark 1' models)

Upper lever – controls the temperature: left for hot, and right for cold, with variable settings in between.

Lower lever – controls the distribution of the air between the windscreen (right) and the car interior (centre position) with variable settings in between. Moving the lever to the left will decrease the quantity of air, until the airflow is shut off with the lever moved fully to the left.

Booster fan switch – The booster fan has two speeds. Press the switch to the first position for low boost, and to the second position for high boost.

The facia side vents supply unheated fresh air from outside the car. The vents are opened by turning the thumb wheels underneath the vents towards the centre of the car. The vents can be swivelled to control the direction of the fresh air. The airflow is increased when the lower heater control lever is moved fully left. Maximum airflow is obtained with the booster fan operating.

The facia centre vent can be opened by lifting the front edge, and supplies a controlled flow of air from the heater. The vent should be closed when demisting the windscreen.

Heating and ventilation controls ('Mark 2' models)

'LOW SERIES' MODELS

Knob **A** – controls the distribution of the air between the windscreen (centre position) and the floor area (fully anti-clockwise when viewed from the passenger's seat), with variable settings in between. Moving the knob clockwise (when viewed from the passenger's seat) from the centre position will decrease the quantity of air, until the airflow is shut off with the knob moved fully clockwise.

Switch **B** – controls the booster fan, which has two speeds. Press the switch to the first position for low boost, and to the second position for high boost.

Knob **C** – controls the temperature: anti-clockwise (when viewed from the passenger's seat) for hot air, and clockwise for cold air, with variable settings in between.

The airflow through the louvred vents at the sides of the facia is controlled independently using the lever below each vent to increase or decrease the amount of air, and by moving the position of the slats to control the direction.

▲ Heating and ventilation controls ('low series Mark 2' models)
A Air distribution control
B Booster fan switch
C Air temperature control

'HIGH SERIES' MODELS

Knob **A** – controls the distribution of the air between the windscreen (centre position) and the floor area (clockwise), with variable positions in between. Moving the knob anti-clockwise from the centre position will reduce the flow of air to the floor area, until the airflow is shut off with the knob moved fully anti-clockwise.

Knob **B** – controls the booster fan, which has three speeds. Turn the knob clockwise to switch on the booster fan. Continue to turn the switch clockwise to increase the fan speed.

Knob **C** – controls the temperature: clockwise for hot air, and anti-clockwise for cold air, with variable settings in between.

The airflow through the louvred vents at the centre and sides of the facia is controlled independently using the lever below each vent

CONTROLS AND EQUIPMENT

to increase or decrease the amount of air, and by moving the position of the slats to control the direction.

▲ Heating and ventilation controls ('high series Mark 2' models)
A Air distribution control
B Booster fan switch
C Air temperature control

Hazard flasher switch

Operates all the direction indicator lights simultaneously, regardless of the position of the ignition switch.

Rear foglight(s) switch

Operates the rear foglight(s), provided that the sidelights or headlights are switched on. A warning light incorporated in the switch indicates that the foglight is switched on.

Heated rear window switch

Operates the heated rear window provided that the ignition is switched on. It's recommended that the device is switched off as soon as demisting is complete, as drain on the battery is very high (some models are fitted with a timer which switches the element off automatically after 10 minutes).

Loudspeaker balance control

Adjusts the sound balance between the front, rear, left, and right of the car by moving the joystick in the required direction.

Instrument illumination control

Adjusts the brightness of the instrument illumination when the exterior lights are switched on.

Remote tailgate release switch

Mounted on the centre console behind the gear lever. Operates provided that the engine is not running.

▲ Remote tailgate release switch location (arrowed)

Bonnet release lever

Mounted under the facia on the passenger's side of the car on 'Mark 1' models, and under the steering column on 'Mark 2' models. Pull the lever to release the bonnet, then lift the **rear** edge of the bonnet and tilt it fully forwards to the vertical position. A safety spring mechanism will hold the bonnet in position: do not push the bonnet beyond its rest position.

When closing the bonnet, allow it to drop the last few inches to the closed position.

▲ Bonnet release lever ('Mark 2' models)

FORD FIESTA

26 CONTROLS AND EQUIPMENT

INTERIOR EQUIPMENT
Cigarette lighter

Operates regardless of the position of the ignition switch. Press the lighter in, then release it, and wait until it pops out ready for use.

OPENING FRONT QUARTERLIGHTS ('MARK 1' MODELS)

On some models, the front quarterlights can be opened after pressing the catch locking tab **(1)** and turning the catch **(2)**.

▲ Pull the lever **(A)** to open the sunroof

▲ Opening front quarterlight ('Mark 1' model)
For **(1)** and **(2)** see text

▲ Push the links **(B)** together ...

Sunroof

The sunroof is of the tilt/removable type. To open the sunroof, pull the lever **(A)** down, then swivel the lever forward and push it up to raise the panel.

To remove the roof panel, pull lever **(A)** down, and push the links **(B)** together to release them from the roof recess. From outside the left-hand side of the car, lift the edge of the panel slightly and release the safety clip **(C)**. Lift the panel to disengage the front guides from the roof, then remove the panel and store it in its protective cover in the boot. The ends of the rubber retaining straps should be pushed into the slots provided in the floor of the boot to hold the panel in place.

When refitting the roof panel, make sure that the front guides engage in the slots in the roof, then engage the safety clip **(C)** and engage the links **(B)** in the roof recess by pushing them together and releasing when in position.

▲ ... and release the safety clip **(C)** to remove the sunroof panel

▲ Sunroof panel protective cover and retaining straps

FORD FIESTA

CONTROLS AND EQUIPMENT

Dipping rear view mirror

Pull back the lever under the mirror to reduce the glare from the lights of following vehicles.

Seat belts

'MARK 1' MODELS

Static seat belts

To wear the belt, remove the tongue from the clip on the door pillar, and push the tongue into the buckle until a click is heard.

To loosen the belt, turn the adjuster at right-angles to the belt, and pull it away from your body. Move the plastic clip to hold the end of the belt.

To tighten the belt, simply pull the webbing through the adjuster.

Inertia reel seat belts

The operation of these seat belts is self-explanatory, noting that the belt should be pulled slowly out of the reel. Ideally, the car should be parked on level ground, otherwise the reel may lock. If the car is parked on a slope, it may not be possible to fit the seat belts until the car is moved onto level ground.

The locking mechanism can be checked by giving the webbing a sharp tug, when the reel should lock.

'MARK 2' MODELS

Front seat belts

The front seat belts are of the inertia reel type, as described for 'Mark 1' models.

On some models, the outer lower end of the belt is free to slide along a steel bar. When putting the belt on, it's essential to ensure that the belt has moved to the front of the bar, and has not been prevented from sliding forward by a rear seat passenger, or by luggage behind the seat. Similarly, when the belt is released, ensure that it returns to the rear of the bar and does not lie in the door aperture.

On some models, the outer lower end of the belt is attached to a bracket which is free to swivel around its mounting bolt. When putting on this type of belt, it's essential to make sure that the bracket is pulled forwards and upwards by the belt, to avoid any slack in the webbing.

Rear seat belts

The side belts are of the inertia reel type as described previously.

Where fitted, the centre rear seat belt is of the lap type, and must be manually adjusted for fit by the wearer. The tongue and the buckle of the lap strap are identified by a white spot, and the lap strap tongue should never be locked into a buckle which does not have this identifying spot. To tighten the belt, pull the webbing through the adjuster, and move the plastic clip to hold the end of the belt. To lengthen the belt, turn the adjuster at right-angles to the belt, then press the tongue into the adjuster housing and pull the webbing through the adjuster.

Front seat adjustment

SEAT SLIDING CONTROL

To move a seat backwards or forwards, pull the lever located under the front of the seat to the right, then slide the seat to the required position. Release the lever, and ensure that the seat has locked in position by gently rocking it backwards and forwards.

SEAT BACK RELEASE

Pull up the lever on the side of the seat, and tilt the seat back forwards to allow access to the rear seats.

SEAT RECLINING CONTROL

Turn the wheel on the bottom edge of the seat to alter the angle of the seat back. The wheel may be located on the inside or the outside edge of the seat, depending on model. If the seat is moved fully forwards, it is possible to recline the seat back further.

Adjustable head restraints

The head restraints can be adjusted for height by pulling up or down (do not force the head restraint beyond the upper stop position). On some models, the angle can be adjusted by tilting backwards or forwards.

Folding rear seats

Either a full-width or a split rear seat back may be fitted, depending on model.

On 'Mark 1' models, the rear parcel shelf

FORD FIESTA

cords must be unhooked from the tailgate, or the parcel shelf must be removed completely, before the seats can be folded.

To fold the seats, pull the relevant release lever(s) on the top of the seat back(s) (pull both levers simultaneously in the case of full-width bench type seats), then push the seat back(s) forward.

▲ *Folding a split rear seat back*

On models fitted with rear seat belts, when returning the seat back(s) to the normal position, hold the top(s) of the rear seat belt(s) forwards to avoid trapping, and make sure that the seat back locking catch(es) engage properly.

Rear parcel shelf

On models with a rear parcel shelf, the shelf can be removed by unhooking the cords from the tailgate, and pressing the retaining catches on the seat back on 'Mark 1' models, or by pulling the shelf rearwards from the supports at the side of the boot on 'Mark 2' models.

EXTERIOR EQUIPMENT

Door locks

'MARK 1' MODELS

To lock a door from outside, close it and turn the key in the lock. Alternatively, with the door open, lift the handle, press the lock interior plunger (located at the top of the door below the window), then release the handle and close the door. Note, however, that this second method of locking can result in the car keys being locked inside the car!

To lock a door from inside, simply press the lock interior plunger.

'MARK 2' MODELS

To lock a door from outside, close it and turn the key in the lock. Alternatively, with the door open, press the lock interior plunger (located at the top of the door below the window), then close the door, making sure that the exterior handle is held up. Note, however, that this second method of locking can result in the car keys being locked inside the car!

To lock a door from inside, simply press the lock interior plunger.

Tailgate lock

To open the tailgate, insert the key in the lock and turn it clockwise. Some models are fitted with a knob which enables the tailgate to be opened without using the key (assuming the tailgate is unlocked) by turning the knob clockwise.

ACCIDENTS AND EMERGENCIES

In the event of an accident, the first priority is safety. This may seem obvious, but in the heat of the moment, it's very easy to overlook certain points which may worsen the situation, or even cause another accident. The course of action to be taken will vary depending on how serious the accident is, and whether anyone is injured, but always try to think clearly, and don't panic.

We don't suggest that you consult this Section at the scene of an accident, but hopefully the following advice will help you to be better prepared to deal with the situation should you be unfortunate enough to appear on the scene of an accident or become involved in one yourself.

HOW TO COPE WITH AN ACCIDENT

Deal with any possible further danger

Further collisions and fire are the dangers in a road accident.

- **If possible warn other traffic**

  Where possible, switch on the car's hazard warning flashers.

  If a warning triangle is carried, position it a reasonable distance away from the scene of the accident, to give approaching drivers sufficient warning to enable them to slow down. Decide from which direction the approaching traffic will have least warning, and position the triangle accordingly.

  If possible, send someone to warn approaching traffic of the danger which exists, and to encourage the traffic to slow down.

- **Switch off the ignition and impose a 'No Smoking' ban**

  This will reduce the possibility of fire, should there be a petrol leak.

Call for assistance

Send someone to call the emergency services (dial 999), and make sure that all the necessary information is provided to the operator. Give the exact location of the accident, and the number of vehicles and if applicable the number of casualties involved. Refer to the *'Motorway breakdowns'* Section on page 38 for details of how to call for assistance on a motorway.

- **Call an ambulance**

  If anyone is seriously injured or trapped in a vehicle.

- **Call the fire brigade**

  If anyone is trapped in a vehicle, or if you think that there is a risk of fire.

- **Call the police**

  If any of the above conditions apply, or if the accident is a hazard to other traffic. In most cases, the accident must be reported to the police within 24 hours even if none of the above conditions apply (refer to *'Requirements of the law in the event of an accident'* on page 32).

Administer first aid

Refer to *'First aid'* on page 30.

Provide your details

If you are involved in the accident as a driver or car owner, provide your personal and vehicle details to anyone having reasonable grounds to ask for them. Also inform the police of the details of the accident as soon as possible, if not already done.

30 ACCIDENTS AND EMERGENCIES

FIRST AID

Always carry a first aid kit
If possible, learn first aid by attending a suitable course – contact the St John Ambulance Association or Brigade, St Andrew's Ambulance Association or the British Red Cross Society in your local area for details. These organisations will also be able to provide further training in heart massage and mouth-to-mouth resuscitation which could enable you to save someone's life in an emergency.

Before proceeding with any kind of first aid treatment, deal with any possible further danger, and call for assistance, as described previously in this Chapter.

To help remember the sequence of action which should be followed when dealing with a seriously injured casualty, use the **ABC** of emergency first aid.

Casualties remaining in vehicles
Any injured people remaining in vehicles should not be moved unless there is a risk of further danger (such as from fire, further collisions etc).

Casualties outside the vehicles.

Recovery position
To prevent the possibility of an unconscious but breathing casualty choking or suffocating, he/she must be placed in the recovery position.

Lie the casualty on his/her stomach with the face to one side and the knee and arm bent as shown.

▲ *The recovery position*

First aid for other injuries
Always treat injuries in the following order of priority:
- Ensure that the casualty has a clear airway. If not treat as described opposite.
- Put any breathing, unconscious casualty into the recovery position as described previously.
- Treat any severe bleeding by applying direct pressure to the wound. Maintain the pressure for at least 10 minutes and apply a suitable clean dressing. If the wound continues to bleed, apply further pressure and dressings **over** that already in place. If bleeding from a limb, and as long as the limb is not broken, lift the affected limb to reduce the bleeding.
- Broken limbs should not be moved, unless the casualty has to be moved in order to avoid further injury. Support the affected limb(s) by placing blankets, bags, etc, alongside.
- Burns should be treated with plenty of cold water as soon as possible. Don't attempt to remove any clothing, but cover with a clean dressing.

Reassurance
The casualty may be in shock, but prompt treatment will minimise this. Reassure the casualty confidently, avoid unnecessary movement, and keep him/her comfortable and warm. Make sure that the casualty is not left alone.

Give the casualty NOTHING to eat, drink or smoke.

FORD FIESTA

ACCIDENTS AND EMERGENCIES

Airway
- **Check for breathing**

Check that the airway is clear. If the casualty is breathing noisily or appears not to be breathing at all, remove any obvious obstruction in the mouth and tilt the head back as far as possible. Breathing may then start.

Breathing
- **If breathing has stopped**

If after clearing the airway the casualty still appears not to be breathing, look to see if the chest or abdomen is moving. Place your ear close to the casualty's mouth to listen and feel for breathing, and look to see if there's any movement of the chest. If you can't detect anything, you must start breathing for the casualty (mouth-to-mouth resuscitation).

To do this, pinch the casualty's nostrils firmly, keep his/her chin raised, and seal your lips around the casualty's mouth. Breathe out through your mouth into the casualty until the chest rises, then remove your mouth and allow the casualty's chest to fall. Give one further breath.

If giving mouth-to-mouth resuscitation to a child, remember that an adult's lungs are significantly larger than those of a child. Care must be taken not to overinflate a child's lungs, as this may cause injury.

- **Check for a pulse at the base of the neck**

If the heart is beating, continue to breathe into the casualty at a rate of one breath every 5 seconds, until he/she is able to breathe unaided, then place the casualty in the recovery position (refer to 'Recovery position' opposite).

Circulation
- **If the heart has stopped**

If there is no pulse at the neck, then external chest compression (heart massage) must be started.

To do this, if the casualty is still in the vehicle, he/she must be removed and laid on his/her back on the ground. Kneel down on one side of the casualty. Feel for the lower half of the casualty's breastbone and place the heel of one of your hands on this part of the bone, keeping your fingers off the casualty's chest. Cover this hand with your other hand, interlocking your fingers. Keeping your arms straight and vertical, press the breastbone down about 4 to 5 centimetres (1½ to 2 inches). The pressure should be smooth but not jerky. Do this 15 times, at the rate of just over once a second. It may help you to count aloud as you press.

Follow this by 2 breaths as described under 'Breathing'.

Repeat the cycle of 2 breaths followed by 15 compressions, twice more, ending with 2 breaths, then re-check the pulse.

- **If there is still no pulse**

Repeat the cycle 9 times and then check the pulse. Continue this procedure until there is a pulse or until professional help arrives.

FORD FIESTA

ACCIDENTS AND EMERGENCIES

REQUIREMENTS OF THE LAW IN THE EVENT OF AN ACCIDENT

The following is taken from the Road Traffic Act of 1988.

If you are involved in an accident – which causes damage or injury to any other person, or another vehicle, or any animal (horse, cattle, ass, mule, sheep, pig, goat or dog) not in your vehicle, or roadside property:

You must
- stop;
- give your own and the vehicle owner's name and address and the registration mark of the vehicle to anyone having reasonable grounds for requiring them;
- if you do not give your name and address to any such person at the time, report the accident to the police as soon as reasonably practicable, and in any case within 24 hours;
- if anyone is injured and you do not produce your certificate of insurance at the time to the police or to anyone who has with reasonable grounds required its production, report the accident to the police as soon as possible, and in any case within 24 hours, and either produce your certificate of insurance to the police when reporting the accident or ensure that it is produced within seven days thereafter at any police station you select.

ESSENTIAL DETAILS TO RECORD AFTER AN ACCIDENT

If you're involved in an accident, note down the following details which will help you to complete the accident report form for your insurance company, and will help you if the police become involved.
- The name and address of the other driver and those of the vehicle owner, if different.
- The name(s) and address(es) of any witness(es) (independent witnesses are particularly important).
- A description of any injury to yourself or others.
- Details of any damage caused to the vehicles involved or other property.
- The name and address of the other driver's insurance company and, if possible, the number of his/her certificate of insurance.
- The registration number of the other vehicle (check this against the tax disc if possible).
- The number of any police officer attending the scene.
- The location, time and date of the accident.
- The speed of the vehicles involved.
- The width of the road, details of road markings and signs, the state of the road surface, and the weather conditions.
- Any marks or debris on the road relevant to the accident.
- A rough sketch showing the vehicle positions before and after the accident. It's helpful to make a note of the vehicle positions in terms of distance from fixed landmarks, such as lamp posts, buildings, etc.
- Whether any of the other vehicle occupants were wearing seat belts.
- If the accident occurred at night or in poor visibility, whether vehicle lights or street lights were switched on.
- If you have a camera, take a picture of the vehicles and the scene.

If the other driver refuses to give you his/her name and address, or if you consider that he/she has committed a criminal offence, inform the police immediately.

FORD FIESTA

ACCIDENT REPORT DETAILS   33

Use this page to record all the relevant details in the event of an accident.
You should transfer all the information recorded on this page onto the Motor Vehicle Accident Report form which you can obtain from your insurance company.

OTHER DRIVER DETAILS

FULL NAME OF DRIVER:

ADDRESS:

POST CODE:

HOME TELEPHONE: WORK TELEPHONE:

DRIVING LICENCE NUMBER: ISSUED:

DATE OF BIRTH: DATE DRIVING TEST PASSED:

TYPE OF LICENCE: (tick box) Full ☐ Provisional ☐ Heavy Goods ☐

PERMITTED GROUPS:

FULL NAME OF OWNER:

ADDRESS:

POST CODE:

TELEPHONE:

INSURANCE COMPANY:

POLICY NUMBER:

AGENT OR BROKER:

TELEPHONE:

OTHER VEHICLE DETAILS

MAKE: MODEL:

YEAR: CC:

REGISTRATION NUMBER:

FORD FIESTA

ACCIDENT REPORT DETAILS

CIRCUMSTANCES OF ACCIDENT

DATE: TIME: am/pm

PLACE: (Street or Road)

TOWN: COUNTY:

SPEED: WERE THE POLICE CALLED? Yes ☐ No ☐

If 'Yes' give details of Police Constabulary concerned:

DETAILS OF WHAT HAPPENED (Please use the page opposite to make a rough sketch map)

WHAT WAS THE WEATHER LIKE?

INDEPENDENT WITNESS 1

TELEPHONE:

INDEPENDENT WITNESS 2

TELEPHONE:

DETAILS OF ANY INJURIES SUSTAINED BY EITHER PARTY:

FORD FIESTA

ACCIDENT REPORT SKETCH MAP   35

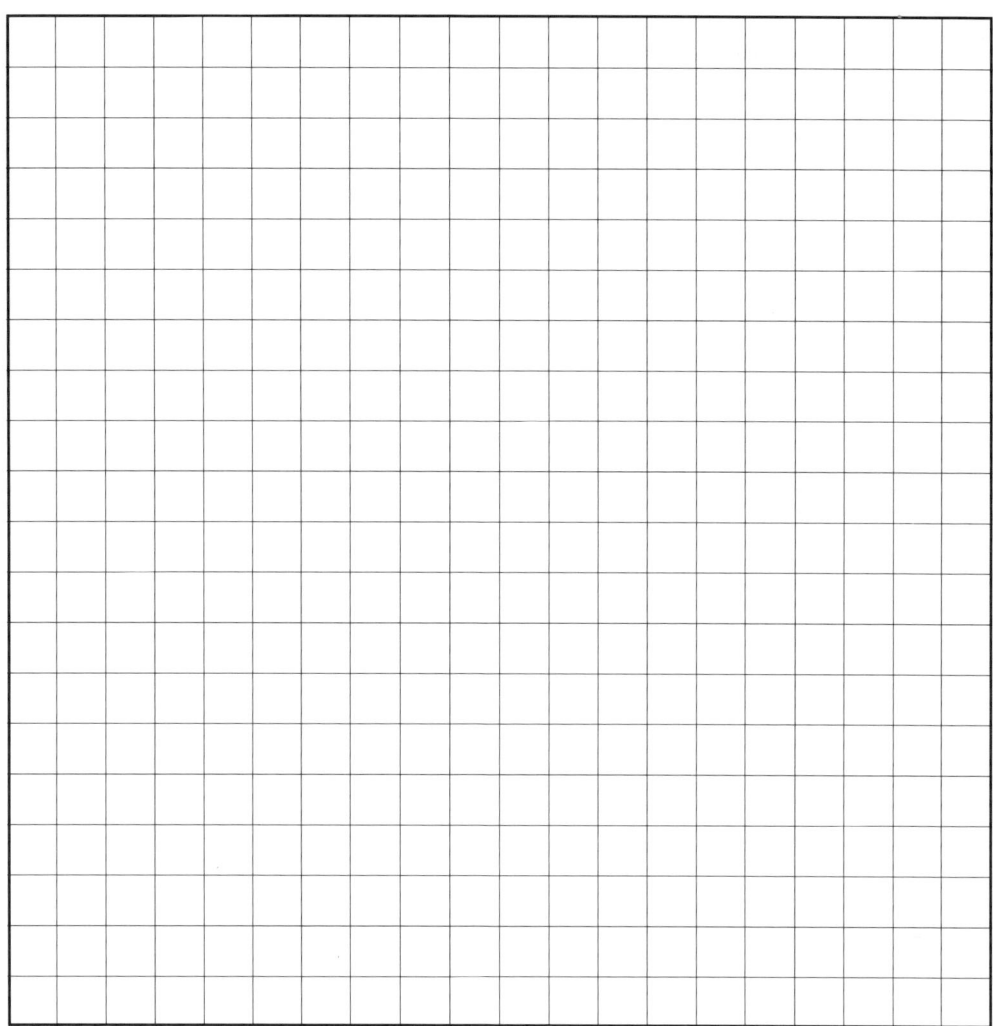

Essential items to include in your sketch map:
- The layout of the road and its approaches.
- The directions and identities of both vehicles.
- Their relative positions at the time of impact.
- The road signs and road markings.
- Names of the Streets and Roads.

FORD FIESTA

ACCIDENTS AND EMERGENCIES

HOW TO COPE WITH A FIRE

Fire is unpredictable, and particularly dangerous when cars are involved because of the presence of petrol, which is highly inflammable.

It's sensible to carry a fire extinguisher on board your car, but bear in mind that the average car fire extinguisher is suitable for use only on the smallest of fires.

In the event of your car catching fire:

- **Switch off the ignition.**
- **Get any passengers and yourself out of the car and well away from danger.**
- **Impose an immediate 'No Smoking' ban.**
- **DO NOT expose yourself, or anyone else, to unnecessary risk in an attempt to control the fire.** Minor fires can be controlled using a suitable extinguisher, but always use an extinguisher at arms-length, and bear in mind that a small vehicle fire can develop into a serious situation without warning.
- **Call for assistance, if necessary** – refer to 'How to cope with an accident' on page 29.

HOW TO COPE WITH A BROKEN WINDSCREEN

There are a number of national specialist companies who offer roadside assistance to drivers who suffer broken windscreens, and it's a good idea to carry the 'phone number of a suitable specialist in your car for use in such situations. Some car insurance policies enable the use of such companies at preferential rates, and it's worth enquiring about this when obtaining insurance quotes.

- **Most of the windscreens fitted to modern cars are of the laminated type.** Laminated windscreens are made from layers of glass and plastic (usually two layers of glass sandwiching a layer of plastic) which prevents the windscreen from shattering, and preserves the driver's vision in the event of the screen suffering an impact.

A sharp impact may crack the outer glass layer, but clear vision is usually maintained. Obviously, if the windscreen is damaged, it should be replaced at the earliest opportunity, but there is no need to curtail your journey unless the damage is particularly serious.

- **Some older cars may be fitted with toughened windscreens.** If a toughened windscreen suffers an impact, the glass will normally stay intact, but severe 'crazing' will usually occur which is likely to seriously affect the driver's vision.

If a toughened windscreen breaks, stop immediately, and seek assistance. **DO NOT** attempt to knock the broken glass out of the windscreen frame, as this is likely to result in injury to yourself, and damage to the car. Driving without a windscreen is extremely dangerous and should not be attempted.

WHAT TO DO IF YOUR CAR IS BROKEN INTO OR VANDALISED

If you are unfortunate enough to have your car broken into or vandalised, do not attempt to drive your car away from the scene unless you are satisfied that no damage has been caused which could affect the car's safety.

Where possible notify the police before moving your car, and inform your insurance company at the earliest opportunity.

BREAKDOWNS

In the unfortunate event of a breakdown, the first priority must always be safety – never risk causing an accident by attempting to move the car or trying to work on it if it's in a dangerous position.

Joining one of the national motoring organisations such as the AA or the RAC can provide you with a recovery and assistance service should you break down. It may also save you money, as local garages often charge high rates to provide a recovery service.

This Section provides advice on the course of action to follow should you break down. There are also Sections covering the procedure to follow for towing, and on how to cope with two of the more common causes of breakdowns – a puncture, and a flat battery.

Breakdowns on an ordinary road (not a motorway)

Most importantly of all, common sense must be used, but the following advice should prove helpful.

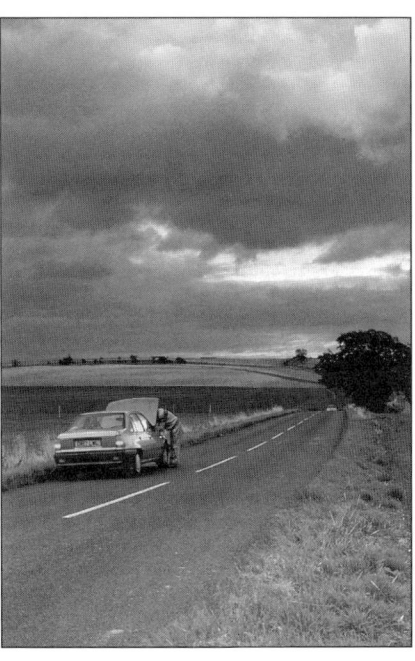

- **Warn approaching traffic,** to minimise the risk of a collision.
  Where possible, switch on the car's hazard warning flashers, and at night leave the sidelights switched on.
  If a warning triangle is carried, position it a reasonable distance away from the scene of the breakdown, to give approaching drivers sufficient warning to enable them to slow down. Decide from which direction the approaching traffic will have least warning, and position the triangle accordingly. Place the warning triangle on the road surface, out from the edge of the road, where it can be easily seen.
  If possible, send someone to warn approaching traffic of the danger which exists, and to encourage the traffic to slow down.
- **Get the passengers out of the car** as a precaution should the car be hit by another vehicle. The passengers should move well away from the car and approaching traffic; up (or down) an embankment, or into a nearby field for example.
- **If possible, move the car to a safe place.** If the car cannot be safely driven or pushed to a place of safety, call a recovery service to provide assistance. *Don't expose yourself or anyone else to unnecessary danger in order to move the car.*
- **If the car can't be safely moved,** turn the steering wheel towards the side of the road. If the car is hit, it will then be pushed into the side of the road, not into the path of traffic.
- **Try to find the cause of the problem.** Only attempt this if the car is in a safe position away from traffic. Refer to *'Fault finding'* on page 121 for details.
- **If necessary call for assistance** from one of the national motoring organisations if you're a member, or from a local garage who can provide a recovery service.

FORD FIESTA

BREAKDOWNS

Motorway breakdowns

In the event of a motorway breakdown, due to the speed and amount of traffic there are a few special points to consider, and the following advice should be followed.

- **If possible, switch on the car's hazard warning flashers, and at night leave the sidelights switched on.**
- **DO NOT open the doors nearest to the carriageway, and DO NOT stand at the rear of the car, or between it and the passing traffic.**
- **Get the car off the carriageway and onto the hard shoulder as quickly as possible, and as far to the left as possible** – never forget the danger from passing traffic. Turn the steering wheel to the left so that the car will be pushed away from the carriageway if hit from behind.
- **If you can't move the car off the carriageway** – get everyone out of the car and well away from the carriageways (up/down an embankment, or into a nearby field for example) as quickly as possible. Never forget the danger from passing traffic. Call for assistance as described below.

  Attempt to warn approaching traffic of your car's presence by signalling from well to the back of the hard shoulder, NOT the carriageway. DO NOT put yourself at risk. Face the approaching traffic, and be prepared to move quickly clear of the hard shoulder onto the verge if any vehicles pull onto the hard shoulder.
- **Get the passengers out of the car** – as a precaution should the car be hit by another vehicle. The passengers should move well away from the car and approaching traffic, up (or down) an embankment, or into a nearby field for example. If animals are being carried, it may be advisable to leave them in the vehicle. In any case, animals and children **must** be kept under tight control.
- **Call for assistance** – using the nearest emergency telephone on your side of the carriageway. NEVER cross the carriageways to use the emergency telephones.

  The direction of the nearest telephone is indicated by an arrow on the marker post behind the hard shoulder (the marker posts are positioned 100 metres/300 feet apart).

  Each telephone is coded with details of its location, and the operator will automatically be informed of your position on the motorway when you make the call.

  Give the operator details of your car type and registration number and, if you're a member of one of the motoring organisations, ask the operator to arrange assistance.
- **Don't leave your car unattended for a long period.**

Changing a wheel

- **First of all, make sure that the car is in a safe position.** It may be safer to drive the car slowly forwards and risk damaging the wheel, than to risk causing an accident.
- **If you're in any doubt** as to whether you can safely change the wheel without putting yourself or other people at risk, call for assistance.
- **All the tools required to change a wheel are kept in the boot.**
  However, if you're carrying any luggage in the boot, you'll need to move it for access to the spare wheel.

  Where applicable, lift up the floor mat, then remove the left-hand cover panel from the floor of the boot to expose the spare wheel. On later models, the cover panel may be secured with a clip. The clip must be lifted and turned, so that it can pass through the slot in the cover panel, before the cover panel can be lifted. Lift the cover panel over the clip, and pull it (towards the back of the car) from the retaining tabs. The spare wheel is secured in place with a bolt which must be unscrewed before the wheel can be lifted out.

  The jack and the wheel brace (a spanner for unscrewing the wheel bolts) may be located under the spare wheel, or under the right-hand boot floor cover panel, depending on model. The wheel brace fits inside the tubular body of the jack. To remove the wheel brace for use, turn the jack handle several turns.
- **Lift the spare wheel and the tools out of the boot,** and place them on the ground. Don't put any luggage back in the boot yet, because the wheel with the punctured tyre will have to be fitted in the spare wheel well.

FORD FIESTA

BREAKDOWNS 39

▲ To remove a boot floor cover panel, lift the retaining clip **(A)** and turn it to align with the slot **(B)** – later models

▲ Jacking points for use with the car jack (arrowed)

▲ Spare wheel, jack and wheel brace locations (spare wheel securing bolt arrowed)

- **Make sure that the handbrake is applied,** and select reverse gear on models with a manual gearbox, or position **P** on models with automatic transmission.
- **Prise off the wheel trim (where applicable),** which covers the wheel bolts on the wheel with the puncture, using the blade-shaped end of the wheel brace.
- **Slacken the four wheel bolts** by at least half a turn, with the wheel on the ground, using the wheel brace. If you carry a length of piping for just such emergencies, slide it over the end of the wheel brace to make it easier to unscrew the bolts.
- **Before jacking the car,** make sure that the ground is solid enough to safely support the jack, and try to find something to chock the wheel diagonally opposite the one to be changed (two large stones or pieces of wood).
- **The jacking points** are located centrally on each side of the car, underneath the door sills. Both wheels on that side of the car will be raised when the jack is used.

Slide the arm of the jack into the jacking point bracket, making sure that the arm slides fully into place, and is not restricted by mud or other debris in the bracket. Unfold the jack handle, then, holding the top of the jack as close as possible to the side of the car, turn the handle until the plate on the bottom of the jack is resting evenly on the ground.

Carry on turning the jack handle slowly to raise the car until the wheel is clear of the ground, making sure that the car doesn't move on the jack.

▲ Using the car jack

FORD FIESTA

40 BREAKDOWNS

Slide the spare wheel under the door sill panel for the time being, on the side nearest the punctured wheel. This will break the car's fall (and minimise the risk of injury to you) if the car slips off the jack or the jack collapses.

- **Unscrew the wheel bolts,** then remove the punctured wheel. Sometimes it may be necessary to tap the wheel in order to remove it, but if this is done, make sure that the car doesn't move on the jack. Put the punctured wheel under the door sill panel, in place of the spare wheel.
- **Fit the spare wheel to the car,** then turn the wheel until the threaded holes in the wheel hub on the car line up with the holes in the wheel.
- **Fit the wheel bolts** and tighten them evenly in a crosswise order, but do not try to fully tighten the bolts with the car still raised. Remove the punctured wheel from under the side of the car, then slowly lower the car to the ground.
- **Remove the jack, and fully tighten the wheel bolts.** There is no need to strain yourself doing this, just make sure that the bolts are tight. If you're unsure of the tightness of the wheel bolts, have them checked at a garage or service station as soon as possible.
- **Refit the wheel trim,** where applicable, making sure that it's correctly aligned and securely fitted (some wheel trims have a hole towards their outer edge which must be aligned with the tyre valve).
- **Stow the punctured wheel** in the boot, together with the jack and wheel brace.
- **Remove the wheel chocks,** and make a final check to ensure that all tools and debris have been cleared from the roadside.
- **Select neutral,** where applicable, before starting up and driving away.

Towing

If your car needs to be towed, or you need to tow another car, special towing eyes are provided at the front and rear of the car.

'Mark 1' models – Where fitted, the towing eyes are positioned on the car underbody. If no towing eyes are fitted, the tow rope must be attached as shown in the accompanying illustration. At the front of the car, the tow

▲ Tow rope attachment points for 'Mark 1' models (arrows indicate towing eyes when fitted)
A Front of car
B Rear of car

rope can be attached to the right- or left-hand suspension tie-rod. At the rear of the car, the tow rope must only be attached to the **left-hand** end of the axle tube, to avoid damage to the suspension or brake components.

'Mark 2' models – The towing eyes are shown in the accompanying illustration. On XR2 models, the front towing eye consists of two hooks; the upper hook is designed for towing the car, and the lower hook **must not** be used for towing. Before using the front towing eye on XR2 models, a plastic cover must be removed from the upper hook by grasping the front edge of the cover and pulling forwards and down. When refitting the cover, locate the lower legs in the grooves provided, and then push the upper edge towards the car.

When being towed, the ignition key should be turned to position **II** (steering lock released, and ignition warning light on). This is necessary for the direction indicators, horn and brake lights to work.

FORD FIESTA

BREAKDOWNS

▲ Towing eye locations (arrowed) – 'Mark 2' models

▲ Removing the front towing eye cover – 'Mark 2' XR2 models
  A Cover lower legs
  B Grooves in front spoiler

Note that on models fitted with a brake vacuum servo (all except some 950 cc engine models), when being towed, the brake servo does not work (because the engine is not running). This means that the brake pedal will have to be pressed harder than usual to operate the brakes, so allowance should be made for greater stopping distances. If the breakdown doesn't prevent the engine from running, the engine could be started and allowed to idle so that the servo operates while towing.

When a car with an automatic transmission is being towed, the gear selector lever should be moved to the 'N' position. To avoid damage to the transmission, do not tow the car any faster than 30 mph (50 km/h) or any further than 30 miles (50 km). Ideally, the car should be towed with the front wheels off the ground, especially if transmission damage is suspected.

An 'On tow' notice should be displayed prominently at the rear of the car on tow to warn other drivers.

Make sure that both drivers know details of the route to be taken before moving off, as it is difficult, and dangerous, to try to communicate once under way.

Before moving away, the tow car should be driven slowly forwards to take up any slack in the tow rope.

The driver of the car on tow should make every effort to keep the tow rope tight at all times, by gently using the brakes if necessary, particularly when driving downhill.

Drive smoothly at all times, particularly when moving away from a standstill. Allow plenty of time to slow down and stop when approaching junctions and traffic queues.

Starting a car with a flat battery

Apart from old age (most batteries should last for at least three years), a battery will normally only go flat if there is a fault in the charging circuit (indicated by a continuously glowing ignition warning light), or when a particular circuit (eg headlights) is left on for a long time with the engine switched off.

To start a car with a flat battery, you can either use a push- or tow-start (models with manual gearbox only), or a set of jump leads.

TOW/PUSH-STARTING

The method used to start the engine is the same whether the car is being towed or being pushed but, if the car is to be towed, first read the previous Section on towing for details of where to connect the tow rope. Proceed as follows:

1 Turn the ignition key to position **II** to switch on the ignition.

2 If the engine is cold, on models with a manual choke, pull out the choke control.

3 Depress the accelerator pedal.

4 Depress the clutch pedal and select third gear. Hold the clutch pedal down.

5 Tow or push the car, and slowly release the clutch pedal. The engine will turn over, and should start (don't worry about the car 'juddering' as the engine turns). If the car is being towed, as soon as the engine starts, depress the clutch pedal and *gently* brake the car so that you don't run into the tow vehicle.

FORD FIESTA

BREAKDOWNS

STARTING USING JUMP LEADS

Note: *Starting using jump leads can be dangerous if done incorrectly. If you're uncertain about the following procedure, it's recommended that you ask someone suitably qualified to do the jump starting for you.*

Before attempting to use jump leads, there are a few important points to note.

First of all, use only proper jump leads which have been specifically designed for the job.

Make sure that the booster battery (the fully-charged battery) is a 12-volt type.

Position the two vehicles close enough together to connect the leads, but **do not** allow the vehicles to touch.

Turn off all electrical circuits.

DO NOT allow the ends of the two jump leads to touch at any time during the following procedure.

Proceed as follows.

1 Open the vehicle bonnets, and connect one end of one of the jump leads (usually the red one) to the positive (+) terminal of the flat battery, and the other end to the positive terminal of the booster battery.

2 Connect one end of the other jump lead to the earth/negative (-) terminal of the booster battery, and connect the other end to a suitable earth point on the car with the flat battery at least 50 cm (18 in) away from the battery (eg a clean, bare metal area on the engine block or body). Make sure that all the lead clips are secure.

3 Start the engine of the vehicle with the booster battery and let it run for a few minutes, then start the engine of the car with the flat battery in the normal way.

4 When the engine is running smoothly, disconnect the jump leads in exactly the reverse order to that in which they were connected.

What to carry in case of a breakdown

Carrying the following items may help to reduce the inconvenience and annoyance caused if you're unlucky enough to break down during a journey:

- Alternator drivebelt (refer to *'Servicing'* on page 95 for details of renewal)
- Spare fuses (refer to *'Bulb, fuse and relay renewal'* on page 114 for details of renewal)
- Set of principal light bulbs (refer to *'Bulb, fuse and relay renewal'* on page 107 for details of renewal)
- Battery jump leads
- Tow-rope (refer to *'Towing'* on page 40 for details of towing procedure)
- Litre of engine oil (refer to *'Regular checks'* on page 66 for details of checking oil level)
- Torch

▲ Jump start lead connections for negative earth vehicles – connect leads in order shown

FORD FIESTA

DRIVING SAFETY

In the UK, more people are injured or killed every year due to road accidents than through any other single cause.

Car manufacturers are paying more attention to passive safety when designing new cars, but drivers must ultimately take active responsibility of ensuring the safety of themselves, their passengers, and just as importantly, other road users.

There will always be accidents on the roads, but taking the time to read the following advice may help you to avoid such a mishap.

The following Sections aim to give advice which will help you to reduce the risk of becoming involved in an incident when driving, not necessarily by changing the way that you drive, but simply by making you aware of some of the potential risks which can easily be avoided. For more information and practical advice on road safety, it is well worth considering taking part in one of the various courses provided by organisations such as RoSPA and the Institute of Advanced Motorists.

BEFORE STARTING A JOURNEY

As a driver, before beginning a journey it's important to make sure that you're comfortable so that you can concentrate on driving without any unnecessary distractions. Spending a few moments carrying out a few simple checks and adjustments will help to make your journey more relaxing, safer and hopefully trouble-free.

Driver comfort

Before driving, make sure that you're comfortable, and that you can operate all the controls easily, particularly if someone else has recently driven the car.

- **Seat**
  Make sure that the seat is positioned a comfortable distance from the steering wheel and the pedals, so that you can operate all the controls comfortably
- **Seat back**
  Make sure that the seat back is adjusted to give your back plenty of support. Your back should rest against the seat, and there should be no need to lean forwards from the seat back. You should be able to reach the steering wheel comfortably, and you should be able to turn the wheel easily without stretching.
- **Head restraint**
  If a head restraint is fitted, it should be adjusted to support your head in the event of an accident. Your head should not rest against the restraint during normal driving and, as a rough guide, the restraint is correctly positioned when its top is in line with your eyes. Similarly, make sure that any passenger head restraints are adjusted correctly if passengers are being carried.
- **Mirrors**
  All the mirrors should be adjusted so that they give a clear view behind the car without the need to move your head unnecessarily.
- **Steering column**
  If the car is fitted with an adjustable steering column, adjust the column so that the steering wheel can be reached comfortably, and the wheel can be turned easily without stretching.

Checks

If you're planning a long journey, refer to *'Regular checks'* on page 65, and make sure that you carry out all the checks mentioned before setting off. Also make sure that you know where the spare wheel and the tools required for wheel changing are located (refer to *'Breakdowns'* on page 38).

DRIVING SAFETY

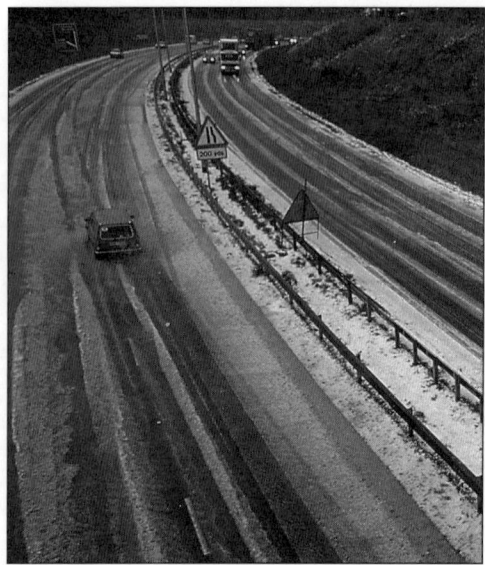

DRIVING IN BAD WEATHER

When driving in bad weather, always be prepared. Bad weather should never take you by surprise. Listen to a weather forecast: you should always be aware of the possibility of poor conditions on your intended journey.

Driving in bad weather requires more concentration, and is more tiring than driving in good weather conditions. Never allow yourself to be distracted by talkative passengers or loud music in the car and, if you feel tired, stop at the next opportunity and take a break.

Always look well ahead, so that you're aware of the condition of the road surface, and any obstacles; **slow down if necessary**.

The following advice is intended to help you to drive more safely in various bad weather conditions. However, above all, use common sense.

Rain

- Use dipped headlights in poor visibility.
- Slow down if visibility is poor or if there is a lot of water on the road surface.
- Keep a safe distance from the vehicle in front (stopping distances are doubled on a wet road surface).
- Be particularly careful after a long period of dry weather. Under these conditions, rain can make the road surface very slippery.
- Don't use rear foglights unless visibility is **seriously** reduced (generally less than 100 metres/300 feet). Foglights can dazzle drivers following behind, especially in motorway spray.

Fog

- Slow down. Fog is deceptive, and you may be driving faster than you think. Fog can also be patchy, and the visibility may suddenly be reduced. Always drive at a speed which allows you to stop in the distance you can see ahead.
- Use dipped headlights. Using main-beam headlights will usually reduce the visibility even further, as the fog will scatter the light.
- Use foglights (where they are fitted) if it's genuinely foggy (generally, where visibility is reduced to less than 100 metres/300 feet), and not just misty.
- Keep a safe distance from the vehicle in front.
- Use your windscreen wipers to clear moisture from the windscreen.

Snow

- Don't start a journey if there is any possibility that conditions may prevent you from reaching your destination. Listen to a weather forecast before setting off.
- Before starting a journey, clear **all** snow from the windscreen, windows and mirrors. Don't just clear a small area big enough to see through.
- Slow down.
- Keep a safe distance from the vehicle in front (stopping distances can be trebled or even quadrupled on a snow-covered or icy surface).
- Drive smoothly and gently. Accelerate gently, steer gently and brake gently.
- Don't brake and steer at the same time – this may cause a skid (refer to 'Skid control' on page 48).
- When moving away from a standstill, or manoeuvring at junctions, etc, in a car with a manual gearbox use the highest possible gear that your car will accept to move away, and change up to a higher gear earlier than

FORD FIESTA

DRIVING SAFETY

usual. In a car with automatic transmission move the selector lever to prevent the transmission from using all the gears (position 'L' – refer to *'Controls and equipment'* on page 23). This will provide more grip and will help to reduce wheelspin.
- Use main roads and motorways where possible. Major roads are likely to have been 'gritted' and are usually cleared before minor roads.
- Use dipped headlights in poor visibility.
- Don't use rear foglights unless visibility is **seriously** reduced (generally less than 100 metres/300 feet) – foglights can dazzle drivers following behind.

Ice and frost
- Follow the advice given for driving in snow.
- Be prepared for 'black ice'. Although you can't see 'black ice', reduced road noise from the tyres will usually tell you it's there.

Severe winter weather
The best advice in severe winter weather is to stay at home but, if you must drive your car, in addition to following the advice given for driving on snow and ice, there are a few additional pieces of advice which you should bear in mind before setting out on a journey.

- Make sure that you tell someone where you're going, and tell them roughly what time you're expecting to arrive at your destination, and what route you're taking.
- Keep a can of de-icer fluid, a scraper, a set of jump leads and a tow rope in the car at all times.
- Carry plenty of warm clothes and blankets.
- Make sure that you have a full tank of petrol before starting your journey. This will allow you to keep the engine running, without fear of running out of petrol, to provide warmth through the car's heating system should you be delayed by conditions, or worse still stuck.
- Pack some pieces of old sacking or similar material, which you can place under the driving wheels to give better traction if you get stuck.
- Carry a shovel, in case you need to dig yourself or someone else out of trouble.

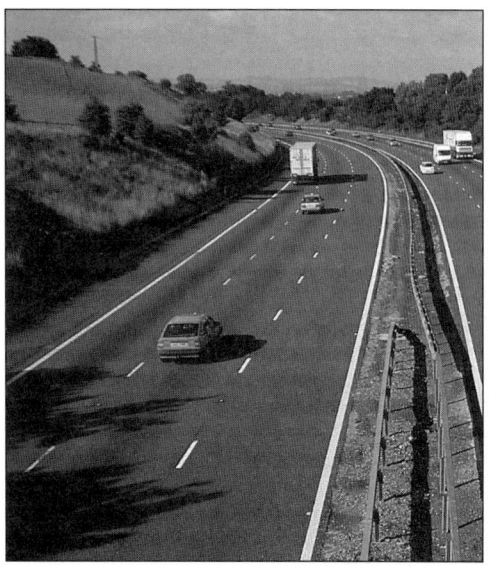

MOTORWAY DRIVING

Motorway driving requires a great deal of concentration and awareness. Motorways are generally busier than ordinary roads, and the traffic moves faster, so there is less time to react to any changes in road conditions ahead.

This Section covers the fundamental rules for safe motorway driving, and will help you to avoid many of the problems which occur on today's busy motorways. Remember that lack of common sense is probably the biggest cause of accidents on motorways.

Full rules and regulations for motorway driving can be found in 'The Highway Code'.

Joining and leaving a motorway
When you join a motorway, you will normally approach from a 'slip-road' on the left. You must give way to traffic already on the motorway. Watch for a safe gap in the traffic, and adjust your speed in the acceleration lane so that when you join the left-hand lane of the motorway you're already travelling at the same speed as the other traffic. Indicate before pulling onto the motorway. After joining the motorway, allow yourself some time to get used to the speed of the traffic before overtaking.

When you leave a motorway, indicate in plenty of time, and reduce your speed before

FORD FIESTA

you enter the slip-road – some slip-roads and link-roads between motorways have sharp bends which can only be taken safely by slowing down. Be very cautious immediately after leaving a motorway, as it can be difficult to judge speed after a long period of fast driving.

Rules for safe motorway driving

- **Concentrate and think ahead** – Always be prepared for the unexpected, and look well ahead. You should be aware of all the traffic in front and behind, not just the vehicle in front of you. If you're concentrating properly, nothing should take you by surprise on a motorway.
- **Always drive at a speed to suit the road conditions** – Don't break the speed limit, especially temporary speed limits which may apply to contra-flow systems or roadworks.
- **Slow down** in bad weather conditions. Driving too fast in fog and motorway spray is a major cause of motorway accidents.
- **Always keep a safe distance from the vehicle in front** – The closer you drive to the vehicle in front, the less chance you have of avoiding an accident if something happens ahead. Remember that you need more space in bad weather conditions.
- **Think** – If the vehicle ahead suddenly stops, do you have enough space to stop without hitting it?
- **Use your mirrors regularly** – It's just as important to be aware of what's happening behind you as it is in front.
- **Always signal your intentions clearly and in plenty of time** – Check your mirrors first, signal *before* you move, and change lanes smoothly and in plenty of time. Don't make any sudden moves which are likely to affect traffic approaching from behind.
- **Keep to the left** – The left-hand lane is *not* the slow lane, and you should drive in this lane whenever possible. You can stay in the middle lane when there are slower vehicles in the left-hand lane, but return to the left-hand lane when you've passed them. The right-hand lane is for overtaking only: it is *not* for driving at a constant high speed. If you use the right-hand lane, move back into the middle lane and then into the left-hand lane

as soon as possible, but without cutting in.
- **Take note of direction signs** – You may need to change lanes to follow a certain route, in which case you should do so in plenty of time.
- **Stop as soon as practicable if you feel tired** – Driving on a motorway when feeling tired can be extremely dangerous. If necessary, wind down the window for fresh air. If you feel tired or drowsy, stop at the next service station, or turn off the motorway at the next junction and take a break before continuing your journey. You must not stop on the hard shoulder of a motorway other than in an emergency.
- **If you break down** – Refer to 'Breakdowns' on page 37.

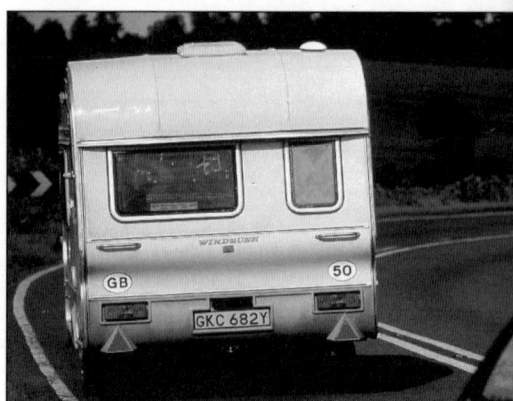

TOWING A TRAILER OR CARAVAN

When towing a trailer or a caravan, there are a few special points to bear in mind. There's more to towing a trailer or caravan than just simply bolting on a towbar and hitching up!

The law

Before using your car for towing, make sure that you're familiar with any special legislation which may apply, particularly if you're travelling abroad.

Make sure that you know the speed limits applicable for towing.

In some countries, including the United

DRIVING SAFETY

Kingdom, there is a legal requirement to have a separate warning light fitted in the car to show that the trailer/caravan direction indicator lights are working.

Always check on current regulations before you travel.

Before starting a journey

Before attempting to use your car for towing, there are one or two points which should be considered.

- **Make sure that your car can cope with the load which you're towing**
- **Don't exceed the maximum trailer or towbar weights for the car** – Refer to 'Dimensions and weights' on page 13.
- **Engine** – Don't put unnecessary strain on your car's engine by trying to tow a very heavy load. Obviously, the smaller the engine, the less load the car can comfortably tow. Also bear in mind that the extra load on the engine when towing means that the engine's cooling system may no longer be adequate. Some manufacturers provide modified cooling system components for towing, such as larger radiators, etc, and if you intend to use your car for towing regularly, it may be worth enquiring about the availability and fitting of these components.
- **Suspension** – Towing puts extra strain on the suspension components, and also affects the handling of the car, since standard suspension components aren't usually designed to cope with towing. Heavy duty rear suspension components are available for most cars, and it's worth considering having these fitted if you intend to use your car for towing regularly. Special stabilisers are also available for fitting to most cars to reduce pitching and 'snaking' movements when towing.
- **Make sure that you can see behind the trailer/caravan using the car's mirrors** – Additional side mirrors with extended arms are available to fit most cars. The mirrors must be fitted with folding arms (so that they fold if hit), and should be adjusted to give a good view to the rear at all times. The mirrors can be removed when the car isn't being used for towing.
- **Make sure that the tyre pressures are correct** – Unless a light, unladen trailer is being towed, the car tyres should be inflated to their 'full load' pressures. Also make sure that where applicable the trailer or caravan tyres are inflated to their recommended pressure.
- **Make sure that the headlights are set correctly** – Check the headlight aim with the trailer/caravan attached, and if necessary adjust the settings to avoid dazzling other drivers.
- **Make sure that the trailer/caravan lights work correctly** – Make sure that the tail lights, brake lights, direction indicator lights and, where applicable, the reversing lights and fog lights all work correctly. Note that the additional load on the direction indicator circuit may cause the lights to flash at a rate slower than the legal limit, and in this case you may need to fit a 'heavy duty' flasher unit.
- **Make sure that the trailer/caravan is correctly loaded** – Refer to the manufacturer's recommendations for details of loading. As a general rule, distribute the weight so that the heaviest items are as near as possible to the trailer/caravan axle. Secure all heavy items so that they can't move. To provide the best possible control and handling of the car when towing, the manufacturers recommend an optimum noseweight for the trailer/caravan when loaded (refer to 'Dimensions and weights' on page 13 for details). The noseweight can be measured using a set of bathroom scales as follows:

Place a stout piece of wood between the trailer/caravan tow hitch cup and the scales platform, and read off the weight with the trailer/caravan level. If necessary, redistribute the load, to arrive as close as possible to the recommended noseweight. **Do not** exceed the maximum recommended noseweight.

Driving tips

Towing a trailer or caravan will obviously affect the handling of the car, and the following advice should be followed when towing.

- **If possible, avoid driving with an unladen car and a loaded trailer/caravan** – The uneven weight distribution will tend to make the car unstable. If this is unavoidable,

DRIVING SAFETY

drive slowly to allow for the instability.
- **Always drive at a safe speed** – The stability of the car and the trailer or caravan decreases as the speed increases. Always drive at a speed which suits the road and weather conditions, and always reduce speed in bad weather and high winds – especially when driving downhill. If the trailer or caravan shows any sign of 'snaking', reduce speed immediately – never try to stop 'snaking' by accelerating.
- **Always brake in good time** – If you're towing a trailer or caravan which has brakes, apply the brakes gently at first, then brake firmly. This will help to prevent the trailer wheels from locking. In cars with a manual gearbox, change into a lower gear before going down a steep hill so that the engine can act as a brake, and similarly, in cars with automatic transmission, move the selector lever to position '2', or '1' in the case of very steep hills.
- **Don't change to a lower gear unnecessarily** – Unless the engine is labouring, stay in as high a gear as possible to keep the engine revs as low as possible. This will help to avoid the engine overheating.

ALCOHOL AND DRIVING

By far the best advice on drinking and driving is **DON'T**.

You may feel fine, but it's a proven fact that even one small alcoholic drink will impair your driving to some extent.

Drinking alcohol has the following effects:

- **Reduces co-ordination**
- **Increases reaction time**
- **Impairs judgement of speed, distance and risk**
- **Encourages a false sense of confidence**

The risk of an accident increases sharply after drinking alcohol, and approximately one third of the total number of people killed in road accidents each year have blood alcohol levels above the legal limit for driving. Remember that you're putting other people as well as yourself at risk if you drive after drinking.

The legal limit for blood alcohol level in the UK is 80 milligrams per 100 millilitres. This doesn't correspond to any particular quantity of drink, as the amount of drink required to reach this level varies from person to person. The driving of many people who feel perfectly sober is seriously affected well below the legal limit.

The penalties for driving over the legal limit are severe, and can mean losing your driving licence, a heavy fine, or imprisonment. It's also important to realise that the laws abroad can be far more severe than in the UK, and some countries have a total ban on driving after drinking alcohol.

- **The safest course of action is not to drink and drive.**

SKID CONTROL

If you drive sensibly with due regard for the road conditions, you should never find yourself in a situation where your car is skidding. The following advice will help you to avoid situations which may cause a car to skid, and explains how to regain control of your car quickly should the need arise.

FORD FIESTA

DRIVING SAFETY 49

A skid is caused by one or a combination of the following:

- **Excessive speed in relation to the road conditions.**
- **Harsh or excessive acceleration.**
- **Sudden or excessive braking.**
- **Coarse or excessive steering.**

The most common basic cause of skidding is rough handling of the car's controls. Therefore the key to safe driving and preventing a skid is smooth, gentle handling of the car and its controls. Try to apply smooth pressure to the brake pedal, accelerator and steering wheel rather than just suddenly moving them a certain distance.

When a car is skidding, the following things happen:

- **The car is out of control. You can take action to regain control, but while skidding, the car can't be fully controlled.**
- **A car with the front wheels skidding can't be steered.**
- **A car with the rear wheels skidding is likely to spin round if any steering lock is applied.**
- **A car with all four wheels skidding will continue in a straight line in the direction it was travelling when the skid started, regardless of which way the car is pointing.**

Prevention is better than cure. When approaching a corner, reduce speed early by smooth progressive braking **in a straight line**. In a car with a manual gearbox, brake to an appropriate speed and select the correct gear for the corner. Turn into the corner early and smoothly, gently increasing the steering lock if necessary, and corner with your foot gently on the accelerator, so that the engine just 'pulls' the car through the corner, maintaining a constant speed. As you leave the corner, accelerate gently and smoothly.

If you follow the above advice, you should avoid finding yourself in a situation where your car is skidding. However, should you be unfortunate enough to experience skidding, the basic rule for skid control is to **remove the cause** of the skid. If you cause a skid by accelerating, braking or steering, **ease off**, and, when the skid stops, re-apply the accelerator, brake or steering control, but this time more gently and smoothly. If the front wheels are skidding, to regain steering control quickly, steer 'into the skid'. This is often misunderstood, and it basically means look in the direction you want to be facing, and steer in that direction – and be prepared to reduce the steering quickly if necessary when the wheels regain their grip.

Several organisations offer courses in skid control, using specially-equipped cars under carefully-controlled conditions. One of these courses could prove to be a very worthwhile investment, possibly helping you to avoid an accident.

ADVICE TO WOMEN DRIVERS

Unfortunately, it's a fact that women are more likely than men to be the subject of unwelcome attention when driving alone.

There's no reason to think that driving alone spells trouble, but it's a good idea to be aware of certain precautions which will help to avoid finding yourself in an unpleasant situation.

This Section aims to give advice which will help to reduce any risks when driving alone, and will help you to deal with the situation, should you be unfortunate enough to find yourself the subject of unwelcome attention.

Several organisations (some local police authorities, AA, etc) run courses especially for women drivers, in which subjects such as basic car maintenance and self defence, etc, are covered.

Sensible precautions
- **The car**

Make sure that you're familiar with your car.

Make sure that your car is regularly serviced. Refer to *'Regular checks'* on page 65, and make sure that all the checks described are carried out regularly. This will minimise the possibility of a breakdown.

Make sure that you know how to change a wheel, and make sure that the car jack is in good condition in case you have a puncture - refer to *'Breakdowns'* on page 38.

FORD FIESTA

DRIVING SAFETY

● The journey

If you're planning to undertake a particularly long or unfamiliar journey, the following advice may be helpful.

If you're travelling on an unfamiliar route, make a few notes before you set off, reminding yourself of the road numbers, where to turn, which junctions to use on motorways, etc. Try to stick to main roads where possible. Always carry a map.

Make sure that you have enough petrol for the journey. If you need to fill up during the journey, make sure that you do so in plenty of time, before the fuel level gets too low, and before petrol stations close if travelling at night.

If you think that your family or friends may be worried, you may want to 'phone someone at your destination before setting off to tell them that you're leaving, and tell them what time you expect to arrive. Similarly, when you arrive, you may want to 'phone someone at your starting point to confirm that you've arrived safely.

● Driving in traffic

Avoid attracting unnecessary attention.

Avoid 'jokey' stickers in car windows, which may encourage unwelcome attention.

Avoid eye contact with 'undesirables'.

If you're being followed, pull over and slow down, but **don't** stop. **Don't** drive to your home, but find a busy and well lit place. If the person following persists, blow your horn and flash your lights to attract attention.

If you're stopped by traffic or another vehicle, lock the doors and close the windows. **Don't** ram the other vehicle - damage to your car might prevent your escape.

● Leaving your car

Don't leave any clues that the car is being driven by a woman; make sure that any 'feminine' items are hidden from sight before leaving your car.

Refer to *'Car crime prevention'* on page 59 and take note of the advice given.

Try to park in a well-lit, preferably busy area.

If you park in a car park, try to park close to an exit, or close to the attendant's station.

Always reverse into a parking space, so that you can drive away quickly if necessary after returning to your car.

Take note of any 'landmarks' so that you can find your car quickly when you return.

Always lock your car.

When you return to your car, walk with a group of people, if possible. Have the keys ready so that you don't have to spend unnecessary time outside your car searching for them, which may attract attention.

Before getting into your car, briefly check for forced entry, and look into the car for any suspicious signs. **Don't** get into the car if you notice anything suspicious.

● Breakdowns

Remember that you're more likely to be injured in an accident than a personal attack.

Refer to *'Breakdowns'* on page 37 and take note of the advice given but, in addition, bear in mind the following advice.

Always walk to 'phone for assistance, **don't** accept a lift. When you 'phone for assistance, mention that you're an unaccompanied woman.

If someone stops while you're 'phoning for help, give the operator details of the other vehicle's registration number and a brief description of the car and driver. If the driver approaches you, tell them that you have passed on his/her details to the police. If the driver's intentions are honourable, your reaction will be understood.

If you decide to stay with, but outside the car, leave the nearside door unlocked and slightly open, so that you can get inside quickly to lock yourself in if necessary.

If you decide to stay inside the car, sit in the passenger seat and lock the door - this will give the appearance that you're accompanied.

When help arrives, ask for some form of identification (even from a policeman) before giving any of your own details.

If someone offers assistance, tell them the police have been informed and are arranging recovery. If you have not yet contacted the police, consider asking the person to do so on your behalf, **but** if you're uncertain about the person's intentions, tell them that the police are aware and ask them to call the police again for you.

FORD FIESTA

DRIVING SAFETY   51

● **Accidents**

Keep calm.

Refer to *'Accidents and emergencies'* on page 29 and take note of the advice given but, in addition, bear in mind the following advice.

If you're bullied or shouted at, lock yourself in your car (if it's safe to do so), and wind the windows up. Communicate through a small gap at the top of the window.

If things get out of hand, refuse to talk to anyone except the police.

Self defence

This is a last resort and should not be used unless all else has failed.

There is no substitute for taking a properly supervised course in self-defence, but we aim to outline a few of the basic moves below.

If you are approached, act in a composed manner and use a soft tone.

In the event of an attack, stay calm - you can still safeguard yourself.

- ● **If you're held by one arm** - Use your free hand to grasp the attacker's thumb and twist the thumb sharply back towards the wrist.
- ● **If you're facing your attacker** - If your shoulders are held, thrust your hands/arms upwards and diagonally outwards to dislodge the attacker's grip.

   Use your knees or the blade of your hand to chop your attacker's groin, your feet to kick the shins, and your fingers to poke the eyes.
- ● **If you're attacked from behind** - Quickly lean forward, twisting your head sideways into the attacker to keep your airway (for breathing) clear. Often this will cause the attacker to lose balance.
- ● **If you've been pulled backwards already** - Turn your head as above, and chop hard with the blade of your hand or a clenched fist to the groin, or an elbow in the stomach.
- ● **If you've been forced to the floor** - Use your feet and legs to kick against the attacker's shins whilst swivelling your body to keep the attacker at bay.

Be ready to run as soon as the attacker's grip is released, and SCREAM!

CHILD SAFETY

Every day, thousands of parents strap their offspring into child car seats, confident that they have done everything possible to protect their loved ones. However, a recent survey has shown that many young children are travelling in car seats which have been incorrectly fitted or are being incorrectly used, making them potentially dangerous should they be involved in an accident. The following advice will help you to make sure that you take every possible precaution to ensure the safety of your children when driving.

How to avoid unnecessary risks

- ● **Never allow young children to travel in a car unrestrained,** even for the shortest of journeys.
- ● **Never carry a child on an adult's lap or in an adult's arms.** Although you may feel that your baby is safer in your arms, this is not the case. An adult holding a child is far more likely to cause injury to the child than to give protection in the event of an accident.

FORD FIESTA

DRIVING SAFETY

- **Never rely solely on an adult seat belt to restrain a child,** and never sit a child on a cushion to enable a seat belt to fit properly.
- **Always strap young children into a properly designed child car seat** when carrying them in a car. The cost of a good car seat is a very small price to pay to save your child's life.

Choosing a child car seat

- It's vitally important to choose the correct type of car seat for your child. A wide range of child car seats is available, and it's worth spending some time looking at the various seats on the market before deciding on the most suitable seat for your particular requirements.
- Although age ranges are often given by the manufacturers, these should be taken as a rough guide. It's the weight of the child which is important; for example, a smaller than average baby could use a baby car seat for longer than a heavy baby of the same age.
- **Never** buy a secondhand child seat. This may sound like a ploy from the manufacturers to sell more seats, but all too often a secondhand seat is sold without the instructions, and sometimes there are parts missing. This often leads to secondhand seats being incorrectly fitted. A secondhand seat may have been damaged or weakened through carelessness or misuse, without necessarily showing any visible signs until it's put to the test in an accident!
- Your baby's first contact with the outside world is often on that first ride home from hospital, so make sure that you're prepared, and buy a car seat before your baby is born.
- Some seats are designed for babies from birth to a weight of around 22 pounds (10 kg), which for most babies is around nine months old. Usually, these first car seats are light and easy to transport through the use of a handle. This means that a sleeping baby can be carried from the car into the house without waking. Some baby seats come complete with a built-in headrest, which is very important for the early weeks when a baby's head needs to be supported.

Certain seats can be fitted so that the baby faces the back of the car (ie rearward facing seats). This may be considered to improve safety, as the baby is supported across the back rather than purely by the harness, if the car is involved in a frontal impact. Combination type seats can then be used forward facing when the child reaches approximately nine months old.

Other types of seat may be designed to accommodate children up to around 40 pounds (18 kg) or approximately four years old.

Alternatively, some seats use the car's seat belts to hold both the child seat and the child, and these seats are obviously easier to fit. Make sure that this type of seat is fitted with a seat-belt lock, so that the seat belt cannot be pulled out of place or slackened. Another advantage of these seats is that they can easily be transferred from car to car.

- Try to choose a seat which has an easily adjustable harness, as this makes it easier to ensure that the harness fits the child securely for each trip. If the harness can be quickly loosened, it makes getting a struggling child in and out a little easier.

Using a child car seat

- Firstly, make absolutely sure that the seat is properly fitted in accordance with the manufacturer's instructions. If you're unsure about any of the fitting procedure, contact the manufacturer for advice.
- To hold a child securely, the harness must be reasonably tight. There should be just enough room to slide your flat hand under the strap. Children may be wearing bulky clothes one day, and thin clothes the next, so it is vital that the harness is adjusted before each journey to ensure a correct fit.
- If the seat is fitted using an adult's inertia reel seat belt, make sure that the seat is held firmly in position, and that the seat belt is securely locked to prevent it from loosening. Also, make sure that the seat belt buckle is not resting on the frame of the child seat. This is because the buckles are not designed to withstand the impact of a heavy child seat and, in an accident, could break open.

FORD FIESTA

DRIVING ABROAD

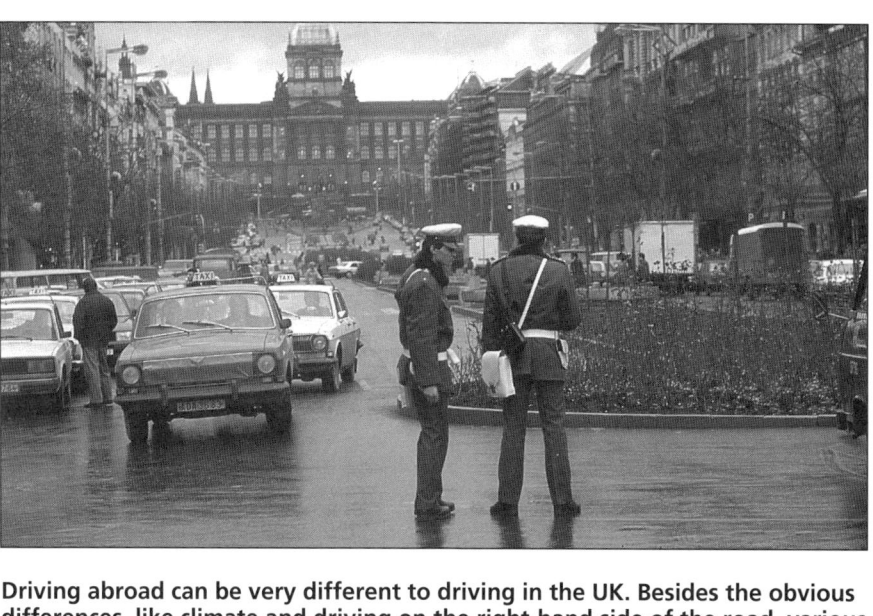

Driving abroad can be very different to driving in the UK. Besides the obvious differences, like climate and driving on the right-hand side of the road, various unfamiliar laws may apply, and it's advisable to prepare yourself and your car as far as possible before travelling.

This Section provides you with a guide which will help you to prepare for driving abroad, and will help you to avoid some of the pitfalls waiting for the unwary.

It may be worthwhile considering hiring a car abroad, rather than driving your own car. In this case, make sure that the insurance cover arranged suits your requirements, and make sure that a damage waiver is included (otherwise you will have to pay for any damage to the hire car).

Insurance
● Motoring
Make sure that you have adequate insurance cover for your car and your luggage.

Check on the legal requirements for insurance in the country you're visiting, and always inform your insurance company that you're taking your car abroad – they will be able to advise you of any special requirements.

Most car insurance policies automatically give the minimum legally-required cover for driving in EC countries, but if you require the same level of cover as you have in the UK, you will normally need to obtain a 'Green Card' (an internationally-recognised certificate of insurance) from your insurance company.

● Medical
It's always advisable to take out medical insurance for the car occupants. Not all countries have a free emergency medical service, and you could find yourself with a large unexpected bill in the event of yourself or one of your passengers being taken ill, or being involved in an accident (in some countries you may even have to pay for an ambulance).

NHS form E111 (available by filling in a form at a Post Office) entitles you to receive

DRIVING ABROAD

the same health care that residents receive in EC countries, but this is by no means comprehensive, and additional insurance cover is usually advisable.

● **Breakdown**

Recovery and breakdown costs can be far higher abroad than in the UK.

Most of the national motoring organisations (such as the AA and RAC) will be able to provide insurance cover which could save you a lot of inconvenience and expense should you be unfortunate enough to break down.

Documents

Always carry your passport, driving licence, car registration document, and insurance certificate (including 'Green Card' and medical insurance, where applicable).

Make sure that all the documents are valid, and that the car's road fund licence (tax) and MOT don't run out while you're abroad.

Before travelling, check with the authorities in the country you're visiting, in case any special documents or permits are required. You may need a visa to visit some countries, and an International Driving Permit (available from the AA or RAC) is sometimes required.

Driving laws

Before driving abroad, make sure that you're familiar with the driving laws in the country you're visiting, as there may be some laws which don't apply in the UK, and the penalties for breaking the law may be severe.

Fit a 'GB' plate to the back of your car, and make sure that it's displayed all the time you're abroad.

In some countries, you're legally required to carry certain items of safety equipment. These can include a first-aid kit, a warning triangle, a fire extinguisher, a set of spare bulbs/fuses, etc.

Remember that if you're visiting a country where you have to drive on the right-hand side of the road, you'll need to fit headlight beam deflectors or shields to avoid dazzling other drivers, or alternatively, it may be possible to have the headlight beams adjusted.

Make sure that you're familiar with the speed limits, and note that in some countries there is an absolute ban on driving after drinking *any* alcohol.

Servicing

Service your car before setting off on your trip, to reduce the possibility of any unexpected breakdowns.

Pay particular attention to the condition of the battery, windscreen wipers and tyres (including the spare), noting that the tyre pressures will probably have to be increased from their normal setting if the car is to be fully loaded. If you're travelling to a cold country, check the condition of the cooling system and the strength of the antifreeze.

Before setting off on your journey, refer to '*Regular checks*' on page 65, and carry out all the checks described. Check that the jack and wheel brace are in place in the car (refer to '*Breakdowns*' on page 38), and that the jack works properly.

Spares

In addition to the items which must be carried by law in the country you're visiting, it's a good idea to carry a few spares which may be difficult to obtain should you need them abroad (one of the national motoring organisations should be able to advise you). For example, you may want to carry clutch and throttle cable repair kits, as right-hand-drive components can be difficult to find outside the UK.

If your car uses a special oil, it's a good idea to take a pack with you.

It's also a good idea to carry a tow rope and a set of jump leads to help you out in case of a breakdown.

Fuel

The type and quality of petrol available varies from country to country, and it's a good idea to check on the availability of the correct petrol type before travelling (again, one of the national motoring organisations will be able to advise you). This is especially important if your car has a catalytic converter, as in this case you must only use unleaded petrol.

Find out what petrol pump markings to look for to give you the correct type and grade of petrol for your car.

Security

A foreign car packed with luggage is an inviting prospect for criminals, so refer to '*Car*

DRIVING ABROAD 55

crime prevention' on page 59, and don't take any chances.

Route planning

It's always a good idea to plan your approximate route before travelling. A vast number of maps and guides are available, or the AA and RAC can provide you with directions to your destination for a modest charge.

Bear in mind that in some countries, you'll have to pay tolls to use certain roads, and this can add unexpectedly to the cost of travelling.

What to carry when driving abroad

The following list provides a guide to the items which it is compulsory to use or carry, or it is strongly recommended that you carry in your car when driving in various European countries.

In the following table **'C'** indicates **'Compulsory'** and **'R'** indicates **'Recommended'**.

COUNTRY	HEADLAMP DEFLECTORS	GB STICKER	SEAT BELTS	WARNING TRIANGLE	FIRE EXTINGUISHER	FIRST AID KIT	SPARE BULBS
Austria	C	C	C	C	-	C	-
Belgium	C	C	C	C	C	-	-
Bulgaria	C	C	C	C	C	C	-
Czechoslovakia	C	C	C	C	-	C	C
Denmark	C	C	C	R	-	-	-
Eire	-	C	C	-	-	-	-
Finland	C	C	C	C	-	-	-
France	C	C	C	C	-	-	R
Germany	C	C	C	C	C	C	R
Greece	C	C	C	C	C	C	-
Holland	C	C	C	C	-	-	C
Hungary	C	C	C	C	-	-	C
Italy	C	C	C	C	-	-	R
Luxembourg	C	C	C	R	-	-	-
Norway	C	C	C	R	-	-	R
Poland	C	C	C	C	-	-	R
Portugal	C	C	C	C	C	-	-
Spain	C	C	C	C	-	-	C
Sweden	C	C	C	R	-	-	-
Switzerland	C	C	C	C	-	-	-
United Kingdom	-	-	C	-	-	-	-
Yugoslavia	C	C	C	C	-	C	C

FORD FIESTA

56 REDUCING THE COST OF MOTORING

FORD FIESTA

REDUCING THE COST OF MOTORING

ECONOMICAL DRIVING

Owning a car is always going to involve some expense, and the running costs can generally be divided into two main areas. The first concerns fixed costs which cannot be avoided, such as car tax and insurance (although obviously you can shop around for the best insurance quote). However, for most owners savings can be made in the second main area of expense, which covers fuel costs and servicing/maintenance bills.

It's surprisingly easy to reduce the amount of fuel used, simply by adapting your driving style to suit the prevailing conditions, and avoiding certain driving habits which tend to increase fuel consumption unnecessarily.

A large proportion of the average servicing bill is made up of garage labour time, money which can be saved by carrying out the work yourself. Refer to *'Servicing'* on page 73 for easy-to-follow instructions on how to carry out most servicing work. It's also worth bearing in mind that maintenance costs can be cut significantly by reducing unnecessary wear-and-tear on the car.

The following advice deals with the most significant causes of high fuel consumption and unnecessary wear-and-tear, and explains how to save money and reduce environmental pollution during everyday driving.

- **Don't warm the engine up with the car standing still** – Engine wear and pollution is at its highest when the engine is warming up, and the engine takes a long time to warm up when running at idle speed with the car stationary. To avoid excessive wear and pollution, drive off as soon as the engine starts, and don't use more 'revs' than necessary.

- **On cars with a manual choke, don't use the choke any longer than necessary** – Push the choke control fully in as soon as the engine will run smoothly without it. When the engine is running with the choke applied, extra fuel is being used and so the fuel consumption is increased, and more pollution is produced.

- **Avoid sudden full throttle acceleration** – Sudden acceleration increases fuel consumption, engine wear and pollution.

- **Don't drive at high engine speeds** – Minimum fuel consumption and pollution is achieved at low engine speeds and in the highest possible gear. Lower engine speed also means less noise and engine wear. For maximum economy, stay in as high a gear as possible for as long as possible, without making the engine labour.

- **Don't always drive at maximum speed** – Fuel consumption, pollution and noise increase rapidly at high speeds. A small reduction in speed (particularly during motorway driving) can significantly lower fuel consumption and pollution.

- **Look well ahead, and drive as smoothly as possible** – By looking well ahead you will be able to react to any change in road conditions in plenty of time, allowing you to brake and accelerate smoothly. Unnecessary or harsh acceleration and braking increases fuel consumption and pollution.

- **Where possible, avoid dense, slow-moving traffic** – In these conditions, more frequent braking, acceleration and gear changing are required, which increase fuel consumption and pollution.

- **Stop the engine during traffic hold-ups** – Obviously if your engine has stopped it doesn't use any fuel and there is no pollution.

FORD FIESTA

REDUCING THE COST OF MOTORING

- **Check the tyre pressures regularly** – Low tyre pressures increase the rolling resistance of the car, and therefore increase fuel consumption, as well as increasing tyre wear and causing handling problems.

- **Don't carry unnecessary luggage** – Weight has a significant effect on fuel consumption, especially in dense traffic where frequent acceleration is required.

- **Don't leave a roof rack fitted when not in use** – The extra air resistance increases fuel consumption.

- **Switch off any unnecessary electrical circuits as soon as possible** – Heated rear windows, foglights, heater blowers, etc, consume a considerable amount of electrical power. The engine must work harder to provide this power by driving the alternator, and the fuel consumption is therefore increased.

- **Check the fuel consumption regularly** – By doing this you will be able to notice any significant increase in fuel consumption, and any problem which may be causing it can be investigated before it develops into anything more serious.

- **Ensure that your car is serviced regularly** – This will ensure that the car operates as efficiently as possible, reducing fuel consumption and pollution.

- **If your car engine is suitable, use unleaded petrol** – This will reduce pollution. If in any doubt as to whether your car's engine is suitable for use with unleaded petrol, seek advice from the car's manufacturer, or from a recognised dealer.

FORD FIESTA

CAR CRIME PREVENTION

Crime against cars is a serious problem, and many owners suffer the consequences of theft or damage every year. The likelihood of your car becoming a target for criminals can be reduced by making it more difficult for your car to be broken into or stolen, and the following advice should prove helpful. Some of the points may seem obvious, but the majority of cars are broken into or stolen in a very short space of time with little force or effort required.

- **Always remove the ignition key** – even in a garage or driveway at home. Always make sure that the steering column lock is engaged before removing the key.

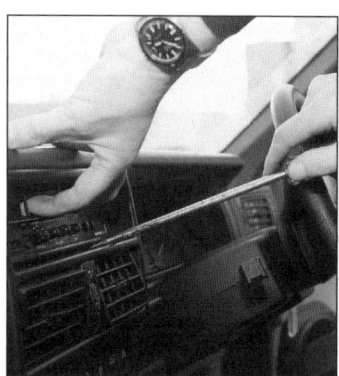

- **Always lock your car** – even in a garage or driveway at home. Where fitted, ensure that 'deadlocks' are engaged before leaving the car and, where applicable, don't forget to lock the fuel filler cover. Make sure that all the windows and the sunroof or folding roof, where applicable, are properly shut. If the car is in a garage, lock the garage. If your car is stolen or broken into while it's unlocked, your insurance company may not pay for the full value of loss or damage.
- **Never leave valuable items on display** – Even if you're only leaving your car for a few minutes, move anything that might be attractive to a thief (even a coat or briefcase) out of sight, and preferably take it with you or lock it safely in the boot. Don't leave valuable items (especially credit cards) in the glovebox. Don't leave your vehicle documents in the car (registration document, MOT certificate, insurance certificate, etc), as they could help a thief to sell it.
- **Park in a visible and (preferably) busy area** – This will deter thieves as they run a greater risk of being caught. If you're parking your car at night, try to leave it in a well-lit area.
- **Put your radio aerial down (where applicable) when parking** – Radio aerials can prove an attractive target for vandals.
- **Protect any in-car entertainment equipment** – The latest security-coded equipment won't work if someone tampers with it and disconnects it from the battery. Some equipment is specially designed so that you can remove the control panel or the entire unit, and take it with you when you leave the car.
- **Fit lockable wheel nuts (or bolts, as applicable)** – if your car is fitted with expensive alloy wheels. Alloy wheels are a favourite target for thieves.
- **Have your car windows etched with the registration number** – This will help to trace your car if it's stolen and the thieves try to change its identity. Other glass components such as sunroofs and headlamps can also be etched if desired. Many garages and specialists can provide this service, and it's possible to buy DIY glass etching kits from motor accessory shops if you prefer to tackle the job yourself.
- **Fit a vehicle immobiliser device** – Many different types are available, but the most common types consist of a substantial metal bar which can be locked in place between the steering wheel and the pedals to prevent the car from being driven.
- **Have an alarm fitted** – Many different types are available, and some are expensive, but they will deter thieves. Some alarms have built-in immobiliser devices to prevent the car from being driven. If you have an alarm fitted, remember to switch it on even if you're only leaving the car for a few minutes.

60   SERVICE SPECIFICATIONS

FORD FIESTA

SERVICE SPECIFICATIONS

Recommended lubricants and fluids

Component/system	Lubricant/fluid type and specification	Duckhams recommendation
ENGINE (1)	Multigrade engine oil, viscosity range SAE 10W/30 to 20W/50, to API SG or better	Duckhams QXR, Hypergrade, or 10W/40 Motor Oil
COOLING SYSTEM 'Mark 1' models (2A) 'Mark 2' models (2B)	55% clean water and 45% antifreeze to Ford specification SSM-97B-9103-A	Duckhams Universal Antifreeze and Summer Coolant
BRAKING SYSTEM (3)	Brake fluid to Ford specification SAM-6C 9103-A	Duckhams Universal Brake and Clutch Fluid
MANUAL GEARBOX (4)	Hypoid gear oil, viscosity SAE 80EP to Ford specification SQM-2C 9008-A	Duckhams Hypoid 80
AUTOMATIC TRANSMISSION (5)	Automatic transmission fluid to Ford specification ESP-M2C 166-H	Duckhams Uni-Matic or Q-Matic

FORD FIESTA

SERVICE SPECIFICATIONS

Lubricant and fluid capacities [Litres (Pints)]

Fuel tank capacity

All models up to August 1984, except 'Mark 2' XR2	34	(7.5 gallons)
'Mark 2' XR2 models, and all models from August 1984	40	(8.8 gallons)

Engine oil

Quantity of oil required to bring level on dipstick from 'MIN' to 'MAX' mark	1.00	(1.75)

Engine oil capacity (for oil and filter change)

OHV type engines	3.25	(5.7)
CVH type engines	3.50	(6.2)

Coolant capacity (for coolant change)

'Mark 1' models:		
950 cc and 1.1 litre engines	5.25	(9.2)
1.3 litre and 1.6 litre engines	6.25	(11.0)
'Mark 2' models:		
950 cc and 1.1 litre engines	5.50	(9.7)
1.3 and 1.4 litre engines	6.30	(11.1)
1.6 litre engine	8.00	(14.1)

Tyre pressures – cold [bars (lbf/in^2)]

Note: *The following is intended as a guide only. Manufacturers frequently change tyre pressure recommendations, and it is suggested that a Ford dealer is consulted for latest recommendations.*

	Front	Rear
Up to 3 occupants:		
135 SR 12 tyres	**1.9** (28)	**1.9** (28)
145 SR 12 and 155 SR 12 tyres	**1.6** (23)	**1.8** (26)
135 SR 13 and 135 R 13-S tyres	**1.8** (26)	**1.8** (26)
155/70 SR 13, 165/65 SR 13, 155/70 R 13-S and 165/65 R 13-S tyres	**1.6** (23)	**1.8** (26)
185/60 SR 13, 185/60 HR 13 and 185/60 R 13-H tyres	**1.8** (26)	**1.8** (26)
Fully laden:		
135 SR 12 tyres	**2.2** (32)	**2.2** (32)
145 SR 12, 155 SR 12 and 155/70 SR 13 tyres	**1.8** (26)	**2.0** (29)
135 SR 13 and 135 R 13-S tyres	**2.6** (38)	**2.8** (41)
165/65 SR 13, 155/70 R 13-S and 165/65 R 13-S tyres	**2.1** (30)	**2.3** (33)
185/60 SR 13, 185/60 HR 13 and 185/60 R 13-H tyres	**2.0** (29)	**2.0** (29)

FORD FIESTA

SERVICE SPECIFICATIONS

Contact breaker points
Points gap [mm (in)]
'Mark 1' models:

Ford distributor	**0.64**	(0.025)
Bosch distributor	**0.45**	(0.018)
'Mark 2' models	**0.40** to **0.50**	(0.016 to 0.020)

Spark plugs

Type	Motorcraft	Champion
950 cc and 1.1 litre engine models	**AGRF 22**	**RS9YCC** or **RS9YC**
1.3 litre OHV engine models with contact breaker points ignition	**AGR 12**	**RN7YCC** or **RN7YC**
1.3 litre OHV engine models with electronic ignition	**AGPR 12C**	**RN7YCC** or **RN7YC**
1.3 litre CVH engine models	**AGPR 22C**	**RC7YCC** or **RC7YC**
1.4 litre engine models	**AGPR 22C1**	**RC7YCC** or **RC7YC**
1.6 litre OHV engine models	**AGPR 12C**	**RC7YCC** or **RC7YC**
1.6 litre CVH engine models	**AGPR 22C**	**RC7YCC** or **RC7YC**

Electrode gap [mm (in)]

Motorcraft plugs:

All 'Mark 1' models	**0.60**	(0.025)
All 'Mark 2' models except 1.4 litre engine models	**0.75**	(0.030)
1.4 litre engine models	**1.00**	(0.039)

Champion plugs:

RS9YCC, RN7YCC and RC7YCC plugs	**0.80**	(0.032)
RS9YC and RC7YC plugs	**0.70**	(0.028)
RN7YC plugs	**0.60**	(0.025)

REGULAR CHECKS

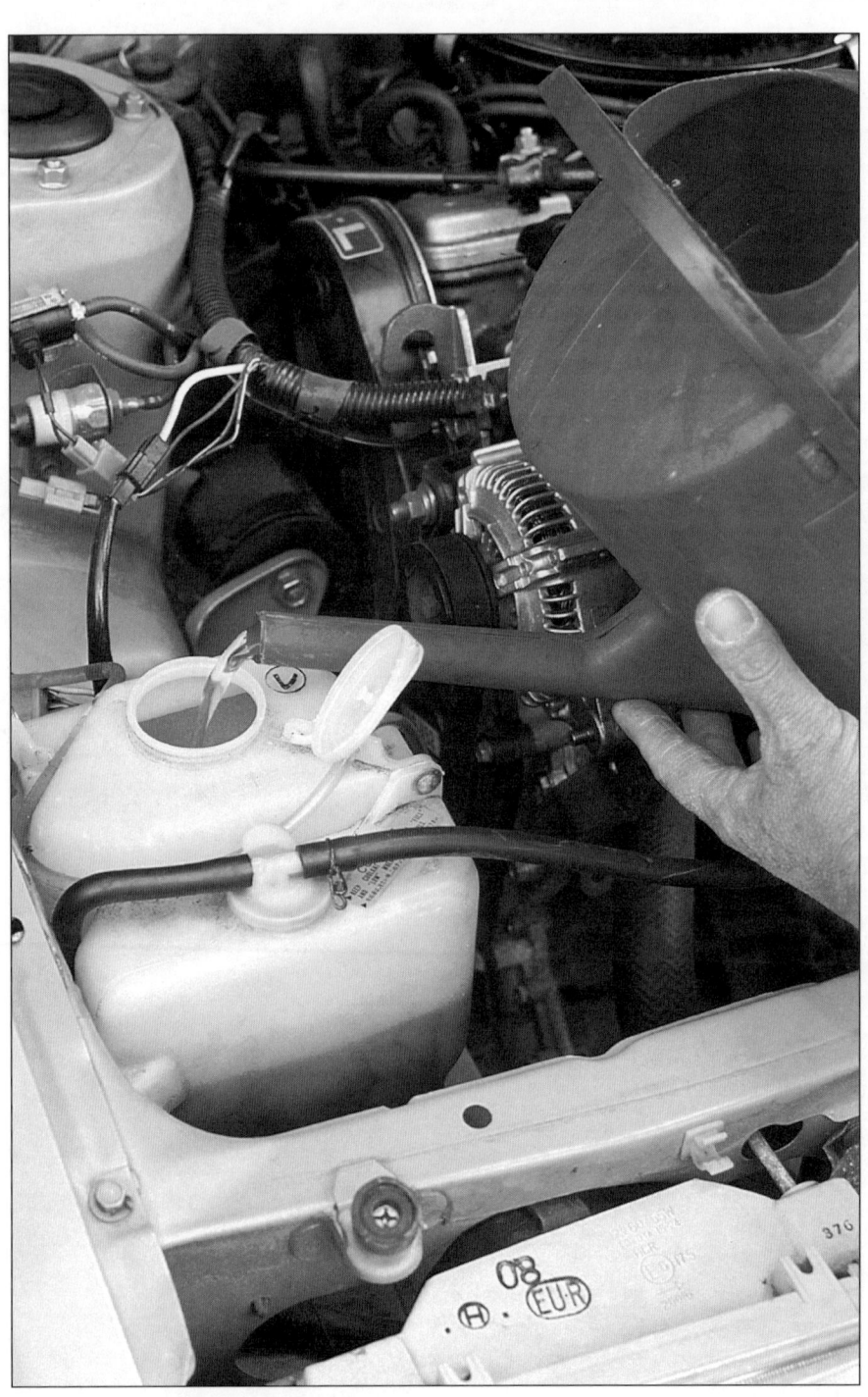

FORD FIESTA

REGULAR CHECKS

65

To ensure that your car is reliable and safe to drive, there are one or two essential checks which are so simple that they're often ignored. These checks only take a few minutes, and could save you a lot of inconvenience and expense. It's a good idea to carry out these checks once a week, and certainly before you start off on a long journey. This Chapter explains how to carry out the checks, and what to do if things aren't quite as they should be.

Whenever you're carrying out checks or servicing jobs, safety must always be the first consideration, and you should bear in mind the advice given in the *'Safety first!'* notes on page 78 before proceeding.

CHECKS

The following checks should be carried out regularly. The checks are explained in more detail in the following pages.

- *Check the oil level*
- *Check the coolant level*
- *Check the brake fluid level*
- *Check the tyres*
- *Check the washer fluid level*
- *Check the battery electrolyte level (where applicable)*
- *Check the wipers and washers*
- *Check the lights and horn*
- *Check for fluid leaks*

ITEMS REQUIRED WHEN CARRYING OUT CHECKS

When carrying out the regular checks, it's a good idea to have the following items close-to-hand. You won't always need all the items, but it's as well to have them available just in case.

- *Small quantity of clean rag* ● *Tyre pressure gauge and foot pump* ● *Small screwdriver (for removing stones from tyres)* ● *Pin (for adjusting washer nozzles)*
- *1.0 litre pack of engine oil (of the correct type – refer to* **'Service specifications'** *on page 61* ● *Small quantity of coolant solution (made up from approximately one-third antifreeze and two-thirds clean water)* ● *Small container of brake fluid (must be an airtight container)* ● *Washer fluid additive* ● *Small quantity of distilled water – for models fitted with batteries which require topping-up*

Note: For details of how to identify engine types, refer to *'Servicing'* on page 83.

REGULAR CHECKS

▲ Typical engine compartment – 'Mark 1' model
  A Brake fluid reservoir
  B Windscreen washer fluid reservoir
  C Cooling system expansion tank
  D Cooling system filler/pressure cap
  E Engine oil filler cap
  F Battery cell cap (remove for electrolyte check)
  G Engine oil level dipstick

▲ Typical engine compartment –
  OHV engine 'Mark 2' model
  A Brake fluid reservoir
  B Battery (maintenance-free type shown)
  C Engine oil filler cap
  D Automatic transmission fluid level dipstick
  E Engine oil level dipstick
  F Cooling system expansion tank and filler/pressure cap
  G Windscreen washer fluid reservoir

▲ Typical engine compartment –
  CVH engine 'Mark 2' model
  A Brake fluid reservoir
  B Battery (maintenance-free type shown)
  C Engine oil filler cap
  D Engine oil level dipstick
  E Cooling system expansion tank and filler/pressure cap
  F Windscreen washer fluid reservoir

Checking oil level

To check the oil level, the car should be standing on level ground with the engine switched off. If the engine has just been run, wait for a few minutes to allow the level to settle before taking a reading.

Open the bonnet, and look for the oil level dipstick, which is at the back of the engine on the right-hand side (when viewed from the driver's seat).

Pull out the dipstick and wipe it clean (with a clean non-fluffy cloth), then slowly push it back into its tube and pull it out again. Note the oil level, which should never be allowed to drop below the 'MIN' mark.

If the oil needs topping-up, pull the oil filler cap from the top of the engine, and top-up to the 'MAX' mark on the dipstick. Note that the amount of oil required to raise the level on the dipstick from the 'MIN' to the 'MAX' mark is approximately 1.0 litre (1.75 pints). Always try to use the same type and make of oil, and take care not to overfill. Mop up any oil which

FORD FIESTA

REGULAR CHECKS 67

▲ *Withdrawing the engine oil level dipstick – CVH engine model*

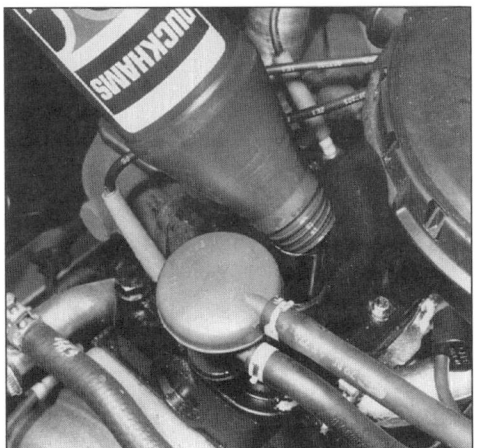

▲ *Topping-up the engine oil level – OHV engine model*

On 'Mark 1' models, the cooling system pressure cap is fitted to the radiator, and when checking the coolant level, the cap should be removed so that the level in the radiator can be checked. The pressure cap should always be removed with the engine cold, but if it does prove necessary to check the level with the engine warm, take care to avoid scalding. Place a thick rag over the cap, and loosen the cap slowly in stages to gradually release the pressure in the system. The level should be within 30 mm (1.2 in) of the flange of the radiator filler neck.

On 'Mark 2' models, the pressure cap is fitted to the expansion tank, and it is not possible to check the coolant level in the radiator.

If topping-up is required, add coolant to the radiator and/or the expansion tank as necessary.

On 'Mark 1' models, to remove the expansion tank cap, twist the cap anti-clockwise, and pull it from the top of the tank.

On 'Mark 2' models, remove the pressure cap from the expansion tank, bearing in mind the precautions described previously for removing the pressure cap on 'Mark 1' models.

Topping-up should always be done with the engine cold, using a mixture of water and antifreeze to the same strength as the mixture already in the system. Although plain water can be used, it's unwise to make it a habit, as the strength of the antifreeze in the main system will gradually be diluted. If there's any might have been spilt, and make sure that the oil filler cap is correctly refitted.

If the oil needs to be topped-up regularly, check for leaks, and if necessary seek advice.

Checking coolant level

The coolant level can be checked visually by checking the level in the expansion tank when the engine is **cold**. When the engine is cold, the level should be up to the 'MAX' mark on the side of the tank. It is normal for the level to be above the 'MAX' mark when the engine is hot.

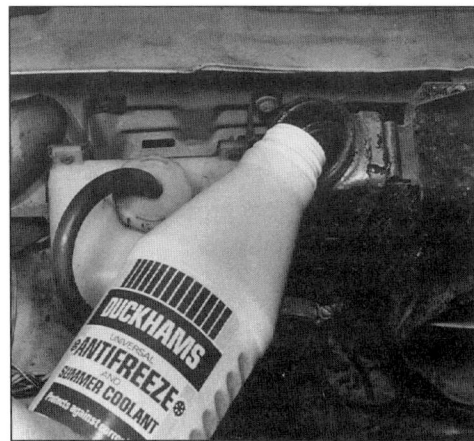

▲ *Topping-up the coolant level in the radiator – 'Mark 1' model*

FORD FIESTA

REGULAR CHECKS

▲ *Topping-up the coolant level in the expansion tank – 'Mark 2' model*

the reservoir, but it will fall slightly as the brake friction material wears. The fluid level must **never** be allowed to drop below the 'MIN' mark.

If topping-up is required, always use the correct type of fluid, which should always be stored in a full, airtight container. Don't top-up using fluid which has been stored in a partly-full container, as it will have absorbed moisture from the air, which can reduce its performance.

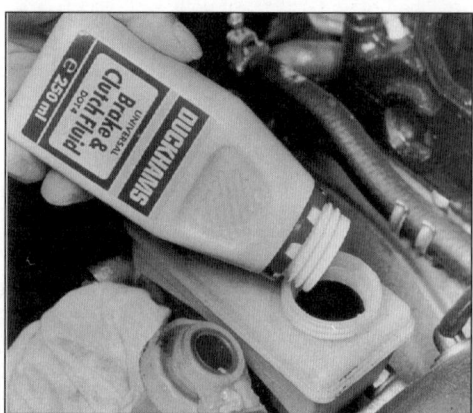

▲ *Topping-up the brake fluid level*

doubt about the strength of the coolant mixture, it's better to add too much antifreeze than not enough.

Normally, topping-up will rarely be required, and if the need for regular topping-up arises, it will probably be due to a leak somewhere in the system. Leaks are most likely to occur from the radiator or the various hoses. If no leaks can be found, it's possible that there's an internal fault in the engine, such as a crack in the cylinder head, or a blown cylinder head gasket, but in this case it's best to seek specialist advice. It's important to note that because of the different types of metals used in the engine, it's vital to use antifreeze with suitable anti-corrosion additives all year round. **Never** use plain water to fill the whole system.

On completion, make sure that the pressure cap is securely refitted. On 'Mark 1' models, make sure that expansion tank cap is securely fitted, and that the end of the overflow tube runs to the bottom of the tank.

Checking brake fluid level

Note: *Refer to* **'Safety first'** *on page 78 for special precautions which should be taken when handling brake fluid.*

Although some models are fitted with a low brake fluid level warning light, the fluid level should be checked visually whenever the oil and coolant levels are checked. The level should be up to the 'MAX' mark on the side of

Topping-up should hardly ever be required unless there's a leak somewhere in the hydraulic system. If a leak is suspected, the car should not be driven until the braking system has been thoroughly checked. **Never** take any risks where brakes are concerned.

Checking tyres

It's extremely important to carry out regular checks on the tyres, to make sure that the pressures are correct, and that the tyres are not damaged. The tyres are the only part of the car in contact with the road, and so their condition will affect the steering and general handling of the car, and therefore its safety.

To check tyre pressures accurately, the tyres must be cold, which means that the car must not have been driven recently. Note that it can make a noticeable difference to the tyre pressures if the car has been standing out in the sun. This can be very noticeable when one side of the car is in the sun, and the other side is in shadow.

FORD FIESTA

REGULAR CHECKS

Note that the recommended tyre pressures vary depending on the size of tyres fitted – refer to *'Service specifications'* on page 62. The size of the tyre (eg 155 SR 13) is clearly marked on the tyre sidewall, although the style of the tyre size marking may vary depending on the tyre manufacturer – consult a Ford dealer or a tyre specialist for further details of tyre size markings and the latest pressure recommendations.

If the tyres are being checked after the car has been driven, for example when filling up with petrol during a journey, the pressures are bound to be higher than specified. It's best to simply check that the pressures in the two front tyres are *equal*, and similarly for the two back tyres (remember that the specified pressure for the rear tyres may be different to the front). **Never** let air out of the tyres of a car which has recently been driven to bring the pressures down to that specified (unless the tyre pressures have been increased for a fully-loaded car, and the load has just been reduced).

When checking the tyre pressures, don't forget to check the spare!

With the tyres properly inflated, run your fingers around the edge of the tyre to check for any cuts and bulges in the tyre walls. Ideally, the car should be jacked up to visually check the tyres all round. Alternatively, the car can be moved backwards and forwards to enable you to check all round the tread. Check all round the tread for cuts, and remove any debris, such as small stones and broken glass, using a screwdriver or similar tool. Any large items such as nails which have penetrated the surface of the tyre should be left in the tread (to identify the location of the damage) until the tyre has been repaired. Refer to *'Breakdowns'* on page 38 for details of how to fit the spare wheel.

Shoulder wear
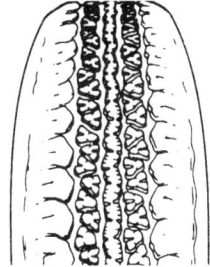

Probable cause:
Underinflation (wear on both sides)
Action: Check and adjust pressure

Probable cause:
Incorrect wheel camber (wear on one side)
Action: Repair or renew suspension parts

Probable cause:
Hard cornering
Action: Reduce speed

Centre wear

Probable cause:
Overinflation
Action: Measure and adjust pressure

Uneven wear

Probable cause:
Incorrect camber or castor
Action: Repair or renew suspension parts

Probable cause:
Malfunctioning suspension
Action: Repair or renew suspension parts

Probable cause
Unbalanced wheel
Action: Balance tyres

Probable cause:
Out-of-round brake disc/drum
Action: Machine or renew disc/drum

Toe wear

Probable cause:
Incorrect toe setting
Action: Adjust front wheel alignment

▲ *Tyre wear patterns and causes*

REGULAR CHECKS

Also check the tyre treads for wear. Uneven wear across one particular tyre may be due to a fault with the suspension or steering. By law, the tread depth must be at least 1.6 mm (0.06 in) throughout a continuous band comprising the central three-quarters of the width of the tyre tread around the full circumference of the tyre.

If you find that any of the tyres is excessively worn or damaged, obtain a new tyre as soon as possible. It's a good idea to try and stick to one type of tyre if possible, rather than mixing several different makes on the car. Generally, you'll find that it's much cheaper to buy tyres from a tyre specialist, rather than an ordinary garage. Make sure that when you buy a new tyre you have the wheel balanced, otherwise you might find that the new tyre causes vibration when driving.

Checking washer fluid level

On 'Mark 1' models, the windscreen washer fluid reservoir is located on the front left-hand side of the engine compartment (when viewed from the driver's seat), in front of the suspension strut top mounting.

On 'Mark 2' models, the windscreen washer fluid reservoir is located towards the front right-hand side of the engine compartment (when viewed from the driver's seat).

Where applicable, the tailgate washer fluid reservoir is located in the boot, under the left-hand boot floor cover panel. Where applicable, lift up the floor mat, then remove the left-hand cover panel from the floor of the boot to expose the reservoir. On later models, the cover panel may be secured with a clip. The clip must be lifted and turned, so that it can pass through the slot in the cover panel, before the cover panel can be lifted. Lift the cover panel over the clip, and pull it (towards the back of the car) from the retaining tabs.

If topping-up is necessary, pull the cap from the reservoir, and top-up as necessary with clean water. A suitable washer fluid additive will keep the glass free from smears, and will prevent the fluid freezing in winter.

Checking battery electrolyte level

From approximately 1982, Ford cars have been fitted with a maintenance-free battery during production. The maintenance-free battery is 'sealed for life', and does not require topping-up with distilled water. A maintenance-free battery is usually clearly marked, and there will be no removable covers to enable topping-up.

On early models with a standard or low maintenance type battery (or on later models where the battery has been replaced by one of this type from another source), the electrolyte level should be checked as follows.

The battery is located at the front right-hand side of the engine compartment (when viewed from the driver's seat) on 'Mark 1' models, and at the front left-hand side of the engine compartment on 'Mark 2' models. Remove the caps or the cover from the top of the battery, and check the level of the electrolyte fluid (with some types of battery you can see the fluid level through the battery case). The level should be just above the tops of the metal plates inside the battery, or in some cases up to a mark on the battery case.

▲ Tailgate washer fluid reservoir location (arrowed)

▲ Removing a cap from the top of a battery to check the electrolyte level

FORD FIESTA

REGULAR CHECKS

If necessary, add distilled water (**not** ordinary tap water) to each cell, to bring the level just above the tops of the battery plates, or up to the marks, as applicable. With some types of battery, distilled water is added to a trough in the top of the battery until all the filling slots are full, and the bottom of the trough is just covered. On completion, refit the caps or cover, and carefully wipe up any drops of water that were spilt.

Topping-up should hardly ever be required, and the need for frequent topping-up indicates that the battery is being overcharged due to a fault in the charging circuit – seek advice from someone suitably qualified if this is the case.

Checking wipers and washers

Check all the wipers and washers to make sure that they're working properly. Don't allow the wipers to work on dry glass for too long, as it will strain the motor and wear out the wiper blades. If necessary, the washer nozzles can be adjusted using a pin inserted into the end of the nozzle. If you're adjusting the windscreen washer nozzles, remember to aim them fairly high on the windscreen, as the airflow will usually deflect the spray down when the car is moving.

Over a period of time, the wiping action of the wiper blades will deteriorate, causing smearing of the glass. The rubber may also crack, particularly at the edges of the blades. When this happens, the blades must be renewed. This is a straightforward job, and is accomplished as follows:

Pull the wiper arm away from the glass, against the spring pressure, until the arm clicks into position. Press the small retaining clip on the blade, then slide the blade out of the hooked part of the arm. The blade can be renewed complete, or the rubber insert can be renewed as follows:

The rubber is held in the arm either by a metal spring clip at one end, or by two metal inserts which run the whole length of the blade.

On models where the rubber is secured by a spring clip, slide back the rubber from the end of the blade, then squeeze the spring clip and pull it from the blade. The rubber can now be slid from the end of the blade.

On models where the rubber is secured by inserts, slide the inserts from either side of the blade and remove the rubber. Note that when refitting the inserts, the cut-outs must face inwards towards the rubber.

Fit the new rubber and/or the blade by reversing the removal operations.

▲ *Press the retaining clip (**A**) and slide the wiper blade from the hooked part of the arm (**B**)*

▲ *Wiper blade rubber retaining spring clip*

▲ *Removing the metal inserts from a wiper blade. Note location of cut-outs*

REGULAR CHECKS

Checking lights and horn

Switch on all the lights and check that they're working. Don't forget to check the direction indicators and the brake lights (the ignition must be switched on to check these), either with the help of an assistant, or by looking for the reflection in a suitable window or door etc.

Check that the direction indicators work with the brake lights on, and that the brake lights work with the tail lights on. Some faults may cause the various rear lights to interact, which can be dangerous, as it may confuse or distract following drivers.

Check the operation of the horn. A short 'blast' should be enough to prove that it at least works, but from time to time check its operation over several seconds.

If any of the bulbs need renewing, refer to 'Bulb, fuse and relay renewal' on page 107.

Any other problems are likely to be caused by a faulty switch, or loose or corroded connections (especially earth connections), and it's best to seek advice to solve these.

Checking for fluid leaks

Check the ground where the car is normally parked for any stains or spots of fluid which may have been caused by fluid leaking from the car.

Open the bonnet, and make a quick check of all the hoses and pipes, and the surfaces of all the components in the engine compartment. If there's any sign of leaking fluid, try to find the source of the leak and seek advice if necessary. It can be difficult to identify leaking fluids, but if there's obviously a major leak, or if you suspect even a slight brake fluid or petrol leak, don't drive the car until the problem has been investigated by someone suitably qualified.

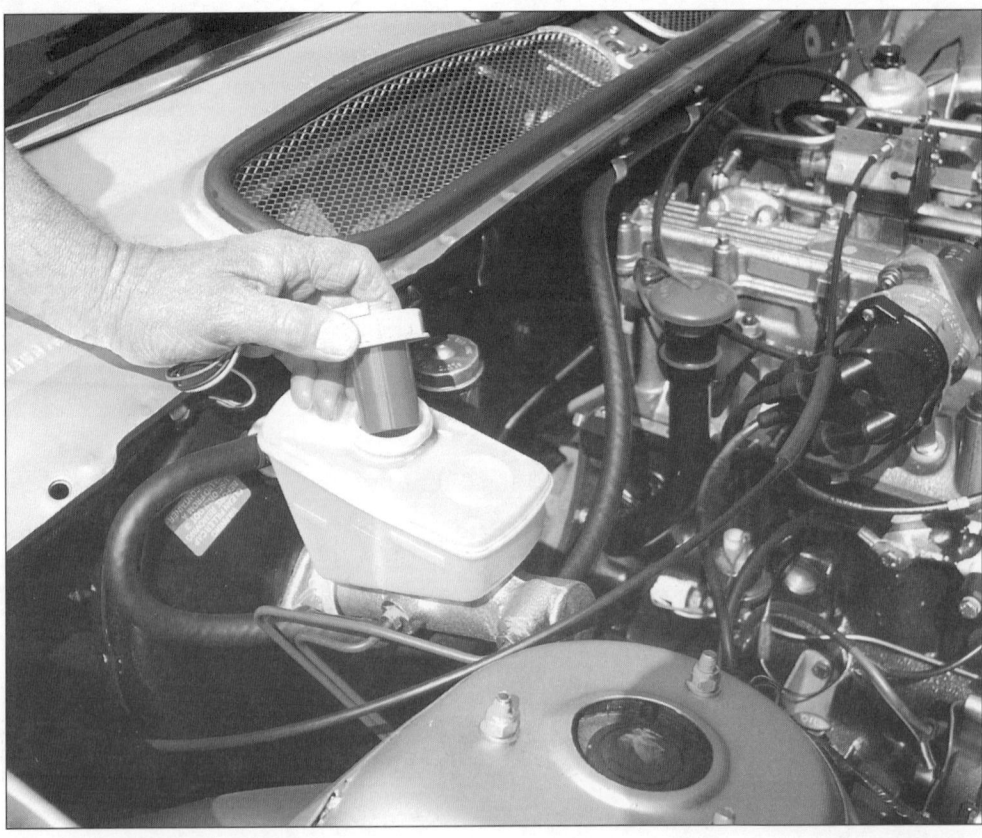

FORD FIESTA

SERVICING

Regular servicing will ensure that your car is reliable and safe to drive, and could save you a lot of money in the long run. Many of the unexpected expenses which can crop up if things go wrong with your car can be avoided by carrying out regular servicing.

A significant proportion of the average garage servicing bill (well over 50% in many cases) is made up of labour costs, so obviously a lot of money can be saved by carrying out the work yourself. You'll find that most of the servicing jobs are very straightforward, and you don't need to be a mechanical genius to 'have a go' yourself. Don't be put off when you look under your bonnet; there are surprisingly few items which require frequent attention, and most of those which do are easily accessible without the need for anything more than basic tools. You'll discover that DIY servicing can be very rewarding, and will help you to understand what makes your car 'tick'.

The following chart lists all the servicing tasks recommended by the car manufacturer, and details of how to carry out the necessary work can be found in the subsequent pages. A few of the jobs require more extensive knowledge, or the use of special tools, and detailed explanation is beyond the scope of this Handbook. These more complicated tasks are identified on the chart, and details can be found in our Owners Workshop Manual for your particular model. Even if you decide to have the work done by a garage, the chart will enable you to check that the necessary work has been carried out.

Whenever you're carrying out servicing, safety must always be the first consideration, and you should read through the *'Safety first!'* notes on page 78 before proceeding any further.

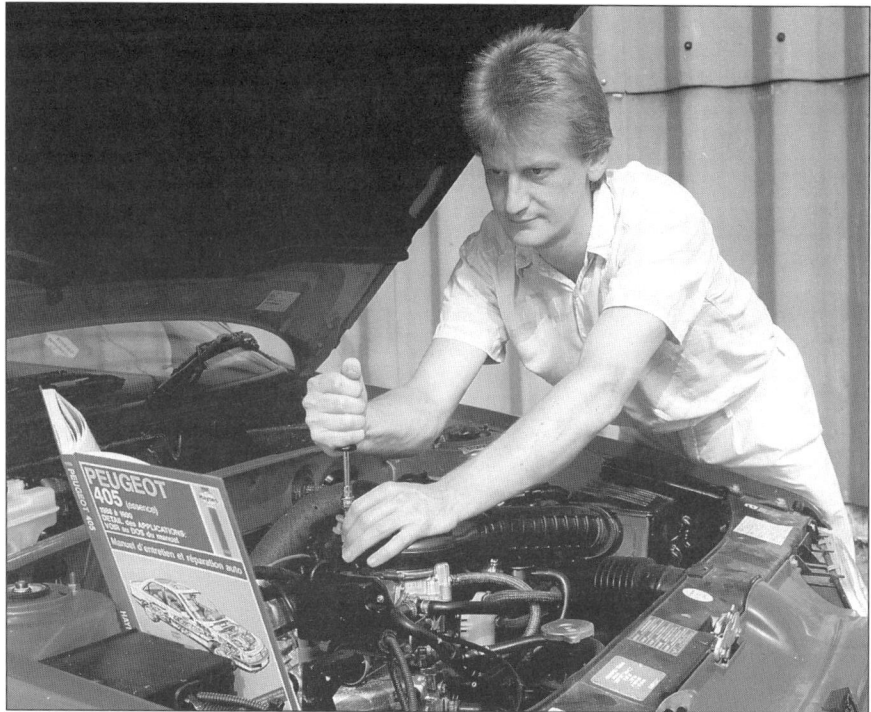

FORD FIESTA

SERVICE SCHEDULE

ENGINE COMPARTMENT
Check all components for fluid leaks, corrosion or deterioration

ENGINE
Check engine oil level

Renew engine oil and filter

On OHV engine models, clean crankcase emission control filter

On CVH engine models, renew crankcase emission control filter

On OHV engine models, check/adjust valve clearances

COOLING SYSTEM
Check coolant level

FUEL AND EXHAUST SYSTEMS
Renew air cleaner filter element

Check/adjust idle speed & mixture

On later OHV engine models, check carburettor throttle damper

IGNITION SYSTEM
Clean distributor cap, coil & HT leads

Check spark plugs

On models with contact breaker distributor, lubricate distributor and check/adjust points gap

Renew spark plugs

On models with contact breaker distributor, renew points

On models with contact breaker distributor, check ignition timing

MANUAL GEARBOX
Check gearbox oil level

AUTOMATIC TRANSMISSION
Check transmission fluid level

BRAKING SYSTEM
Check brake fluid level

Check operation of low fluid level warning system

Check handbrake

Check front brake pads

Check rear brake shoes

Renew brake hydraulic fluid

FORD FIESTA

SERVICE SCHEDULE 75

Regular Weekly	6000 miles (10 000 km) 6 months	12 000 miles (20 000 km) 1 year	24 000 miles (40 000 km) 2 years	36 000 miles (60 000 km) 3 years
	■	■	■	■
■	■	■	■	■
	■	■	■	■
			■	
			■	
		☐	☐	☐
■	■	■	■	■
			■	
	☐	☐	☐	☐
	☐	☐	☐	☐
	■	■	■	■
	■	■	■	■
	■	■	■	■
		■	■	■
		■	■	■
	☐	☐	☐	☐
		■	■	■
		■	■	■
■	■	■	■	■
	■	■	■	■
		■	■	■
	☐	☐	☐	☐
	☐	☐	☐	☐
				☐

Items marked ☐ are considered to be beyond the scope of this Handbook, and details can be found in the Owners Workshop Manual for your car (OWM 334 for Fiesta 'Mark 1' models, or OWM 1030 for Fiesta 'Mark 2' models).

FORD FIESTA

SERVICE SCHEDULE

SUSPENSION AND STEERING
Check tyres for condition and pressure

Check tightness of roadwheel bolts

Check steering components

Check front wheel alignment

INTERIOR AND BODYWORK
Check seat belts

Lubricate hinges & locks

Inspect paintwork & bodywork

ELECTRICAL SYSTEM
Check washer fluid level

Check battery electrolyte level (where applicable)

Check wipers and washers

Check operation of all lights and electrical equipment

Check battery lead connections

Check alternator drivebelt

CAR UNDERSIDE
Check all components for fluid leaks, corrosion or deterioration

ROAD TEST
Check instruments and electrical equipment

Check steering, suspension and general handling

Check engine, clutch (where applicable), gearbox and driveshafts

Check braking system

Note: *In addition to the tasks listed in the table, carry out the following:*

Every 24 000 miles (40 000 km) or 1 year,
renew the automatic transmission fluid (where applicable) – refer to a Ford dealer.
Every 36 000 miles (60 000 km),
on CVH engines, renew the timing belt – refer to Owners Workshop Manual.
Every 36 000 miles (60 000 km) or 2 years,
renew the coolant – refer to **'Service tasks'** on page 98.

FORD FIESTA

SERVICE SCHEDULE

Regular	6000 miles (10 000 km)	12 000 miles (20 000 km)	24 000 miles (40 000 km)	36 000 miles (60 000 km)
Weekly	6 months	1 year	2 years	3 years
■	■	■	■	■
	■	■	■	■
		☐	☐	☐
				☐
	■	■	■	■
	■	■	■	■
	■	■	■	■
■	■	■	■	■
■	■	■	■	■
■	■	■	■	■
■	■	■	■	■
		■	■	■
		■	■	■
	■	■	■	■
		■	■	■
		■	■	■
		■	■	■
		■	■	■

Items marked ☐ are considered to be beyond the scope of this Handbook, and details can be found in the Owners Workshop Manual for your car (OWM 334 for Fiesta 'Mark 1' models, or OWM 1030 for Fiesta 'Mark 2' models).

FORD FIESTA

SERVICING

SAFETY FIRST!

Professional motor mechanics are trained in safe working procedures. No matter how enthusiastic you may be about getting on with the job you have planned, take time to read this Section. Don't risk an injury by failing to follow the simple safety rules explained here. The following is a guide to encourage you to be 'safety conscious' as you work on your car.

Essential points

ALWAYS use a safe system of supporting your car when working underneath it. The car's own jack (or a single hydraulic jack) is never sufficient. Once you've raised the car you should use a safe means of holding it up. Axle stands, ramps or substantial wooden blocks are good; concrete blocks and bricks may crack or disintegrate and should not be used. Make sure that you locate the supports where you know the car won't collapse and where they can't slip.

ALWAYS loosen wheel nuts and other 'high torque' nuts or bolts (as applicable for the task to be carried out) *before* jacking up your car, or it may slip and fall.

ALWAYS make sure that your car is in **Neutral** or in **Park** (in the case of automatic transmission), and that the handbrake is securely on before trying to start the engine.

UNLESS the engine is cold, **ALWAYS** cover the cooling system's pressure cap (radiator or expansion tank cap) with several thicknesses of cloth before trying to remove it. Release the pressure slowly or the coolant may escape suddenly and scald you.

ALWAYS wait until the engine has cooled down before draining the engine (and/or transmission) oil, or the coolant. Oil or coolant that is very hot may scald you. If you can touch the engine sump/transmission/radiator (as applicable) without discomfort, the engine has probably cooled sufficiently to avoid scalding.

ALWAYS allow the engine to cool before working on it. Many parts of the engine become very hot in normal operation and you may burn yourself badly.

ALWAYS keep brake fluid, antifreeze and other similar liquids away from your car's paintwork as it may damage the finish (or may remove the paint altogether).

ALWAYS keep toxic fluids, such as fuel, brake and transmission fluids and antifreeze off your skin, and never syphon them by mouth.

ALWAYS wear a mask when doing any dusty work, or where spraying is involved, especially when doing body repair and brake jobs.

ALWAYS clean up oil and grease spills – someone may slip and be injured.

ALWAYS use the right tool for the job. Spanners and screwdrivers which don't fit properly are likely to slip and cause injury.

ALWAYS get help to lift and handle heavy parts – it's never worth risking an injury.

ALWAYS take time over the jobs you take on. Plan out what you must do, and make sure that you have the right tools and spare parts. Follow the recommended steps, and check over the job once you're finished. Leave 'short-cuts' to the experts.

ALWAYS wear eye protection when using power tools such as drills, sanders and grinders, when using a hammer, and when working beneath your car.

ALWAYS use a barrier cream on your hands when doing dirty jobs, especially when in contact with fuel, oils, greases, and brake and transmission fluids. It will help protect your skin against infection and will make the dirt easier to remove later. Make sure that your hands aren't slippery. The use of a suitable specialist hand cleaner will make dirt and grease easier to remove without causing infection or damage to skin. Long term or regular contact with used oils and fuel can be a health hazard.

ALWAYS keep loose clothing and long hair out of the way of any moving parts.

SERVICING

ALWAYS take off rings, watches, bracelets and neck chains before starting work on your car, especially when working on the electrical system.

ALWAYS make sure that any jacking equipment or lifting tackle you use has a safe working load which will cope with what you intend to do, and use the equipment exactly as recommended by the manufacturers.

ALWAYS keep your work area tidy; someone may trip or slip on articles left lying around and be injured.

ALWAYS get someone to check up on you from time to time if you're working on the car alone, to make sure that you're alright.

ALWAYS do the work in a logical order, check that you've put things back together properly, and that everything is tightened as it should be.

ALWAYS keep children and animals away from an unattended car, and from the area where you're working.

ALWAYS park cars with catalytic converters away from materials which may burn, such as dry grass, oily rags, etc if the engine has recently been run. Catalytic converters reach extremely high temperatures, and any such materials close by may catch fire.

REMEMBER that your safety, and that of your car and other people rests with you. If you are in any doubt about anything, get specialist advice and help right away.

IF in spite of these precautions you injure yourself – get medical help immediately.

Asbestos

Some parts of your car, such as brake pads and linings, clutch linings and gaskets contain asbestos, and you should take appropriate precautions to avoid inhaling their dust (which is hazardous to health) when working with them. If in doubt, assume that they **do** contain asbestos.

Fire

Remember at all times that petrol is highly flammable. Never smoke, or have any kind of naked flame around, when working on the car. But the risk doesn't end there – a spark caused by an electrical short-circuit, by two metal surfaces contacting each other, by careless use of tools, or even by static electricity built up in your body under certain conditions, can ignite petrol vapour, which in a confined space is highly explosive.

If a fuel leak is suspected, try to find the cause and seek advice as soon as possible, and never risk fuel leaking onto a hot engine or exhaust. Catalytic converters (where fitted) run at extremely high temperatures, and therefore can be an additional fire hazard – observe the precautions outlined previously in this Section.

It's recommended that a fire extinguisher of a type suitable for fuel and electrical fires is kept handy in the garage or workplace at all times.

Never try to extinguish a fuel or electrical fire with water.

Fumes

Certain fumes are highly toxic and can quickly cause unconsciousness and even death if inhaled to any extent, especially if inhalation takes place through a lighted cigarette or pipe. Petrol vapour comes into this category, as do the vapours from certain solvents such as trichloroethylene. Any draining or pouring of such fluids should be done in a well-ventilated area.

When using cleaning fluids and solvents, read the instructions carefully. Never use materials from unmarked containers – they may give off poisonous vapours.

Never run a car engine in an enclosed space such as a garage. Exhaust fumes contain carbon monoxide which is extremely poisonous; if you need to run the engine, always do so in the open air, or at least have the rear of the car outside the workplace. Although cars fitted with catalytic converters produce far less toxic exhaust gases, the above precautions should still be observed.

If you're fortunate enough to have the use of an inspection pit, never run the engine while the car is standing over it; the fumes, being heavier than air, will concentrate in the pit with possibly lethal results.

The battery

Never short across the two poles of the battery! (A battery is shorted by connecting the two terminals directly to each other, and this can happen accidentally when working under the bonnet with metal tools.) The heavy

FORD FIESTA

SERVICING

discharge caused will create 'gassing' of hydrogen from the battery and this is highly explosive. Shorting across a battery may also cause sparks, and the combination can cause the battery to explode with potentially lethal results. A conventional battery will normally be giving off a certain amount of hydrogen all the time, so for the same reasons given above, never cause a spark or allow a naked flame close to it.

Batteries which are sealed for life require special precautions which are normally outlined on a label attached to the battery. Such precautions usually relate to battery charging and jump starting from another vehicle (for details of jump starting refer to *'Breakdowns'* on page 42).

If possible, loosen the filler plugs or cover when charging the battery from an external source. Don't charge at an excessive rate, or the battery may burst. Special care should be taken with the use of high charge-rate boost chargers to prevent the battery from overheating.

Take care when topping-up and when carrying the battery. The battery contains dilute sulphuric acid which is very corrosive. It will burn and may cause long-term damage if in contact with skin, eyes or clothes. Similarly, corrosive deposits around the battery terminals may be harmful.

Always wear eye protection when cleaning the battery to prevent the corrosive deposits from entering your eyes.

Mains electricity and electrical equipment

When using any electrical equipment which works from the mains, such as an electric drill or inspection light, always ensure that the appliance is correctly connected to its plug and that, where necessary, it's properly earthed. Always use an RCD (Residual Current Device, a safety device incorporating a circuit breaker) when using mains electrical equipment. Don't use such appliances in damp conditions, and take care not to create a spark or apply excessive heat close to fuel or fuel vapour. Also make sure that the appliances meet the relevant national safety standards.

Ignition HT voltage

A severe electric shock can result from touching certain parts of the ignition system, such as the spark plug HT leads, when the engine is running or being cranked, particularly if components are damp or the insulation is faulty. Where an electronic ignition system is fitted, the HT voltage is much higher than that used in a contact breaker system, and could prove fatal, especially to wearers of cardiac pacemakers.

Disposing of used engine oil

Used engine oil is a hazard to health and the environment, and should be disposed of safely and cleanly. Most local authorities provide a disposal site which will have a special tank for waste oil. If in doubt, contact your local council for advice on where you can dispose of oil safely – there may be a local garage who will allow you to use their specialised oil disposal tank free of charge. Remember that it's not just the oil which causes a problem, but empty containers, old oil filters, and oily rags, so these should be taken to your local disposal site too.

Do not under any circumstances pour engine oil down a drain, or bury it in the ground.

Jacking and vehicle support

The jack provided with the car is designed for emergency wheel changing, and should not be used for servicing and overhaul work. Instead, a more substantial workshop jack (trolley jack or similar) should be used. Whichever type is used, it's essential that additional safety support is provided by means of axle stands designed for this purpose. Never use makeshift means such as narrow wooden blocks or piles of house bricks, as these can easily topple or, in the case of bricks, disintegrate under the weight of the car. Further information on the correct positioning of the jack and axle stands is provided at the end of this Section.

If you don't need to remove the wheels, the use of drive-on ramps is recommended. Ensure that the ramps are correctly aligned with the wheels, and that the car is not driven too far along them so that it promptly falls off the other end or tips the ramps.

FORD FIESTA

SERVICING

JACK AND AXLE STAND POSITIONS

When using a trolley jack or other type of workshop jack to raise the front of the car, the jack head should be placed under the front longitudinal engine/gearbox support member on all 'Mark 1' models and 'Mark 2' XR2 models, or under the centre of the front body crossmember on all 'Mark 2' models except for the XR2.

To raise the rear of the car, the head of the jack should be placed under the centre of the rear axle tube. Great care must be taken not to damage the handbrake cable or the suspension Panhard rod, and to avoid this a wooden support should be made up to act as an insulator between the head of the jack and the axle tube. The support should have a V-shaped cut-out to ensure that it locates securely on the axle tube, and suitable dimensions for a support block are given in the accompanying illustration.

The sides of the car can be jacked up by positioning the jack head under the central side jacking points which are used with the car jack when wheel-changing.

Axle stands should only be located under the double-skinned sections of the side or chassis members as shown in the accompanying illustrations. Note that when using axle stands to support the rear of the car, the axle stands must be located forward of the suspension trailing arms, **not** under them.

▲ When axle stands are used at the rear of the car, they must be located forward of the trailing arms **(A)**, not under them **(B)**

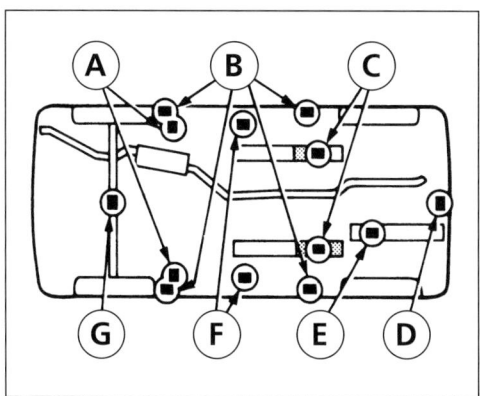

▲ Jacking and support points on car underside
  A, B and **C** Axle stand positions
  D Front jacking point for use with workshop jack (all models except 'Mark 2' XR2)
  E Front jacking point for use with workshop jack ('Mark 2' XR2 models)
  F Side jacking points for use with workshop jack
  G Rear jacking point for use with workshop jack
  Note: At points **(C)** it is important that the car is only supported on the double skinned area of the chassis member

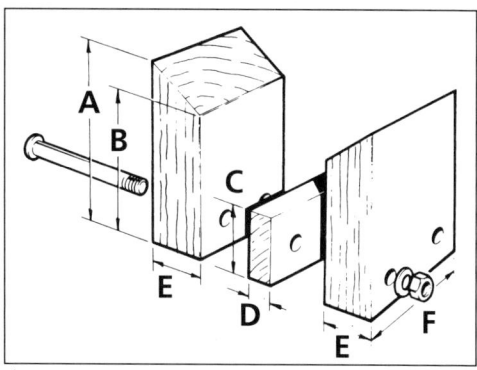

▲ Dimensions for rear axle tube support block
  A 133 mm (5.25 in)
  B 108 mm (4.25 in)
  C 51 mm (2.0 in)
  D 13 mm (0.5 in)
  E 38 mm (1.5 in)
  F 127 mm (5.0 in)

FORD FIESTA

SERVICING

▲ Underbonnet component locations on a 'Mark 2' OHV engine model (air cleaner removed for clarity)

1. Battery earth terminal
2. Radiator cooling fan
3. Engine oil filler cap
4. Carburettor
5. Number 1 spark plug HT lead
6. Alternator
7. Windscreen washer fluid reservoir
8. Brake fluid reservoir
9. Ignition coil
10. Oil filter
11. Vehicle Identification Number (VIN) plate
12. Distributor cap and HT leads
13. Thermostat housing
14. Engine oil level dipstick
15. Cooling system pressure cap

BUYING SPARE PARTS

When buying spare parts such as an oil filter, or an air filter, make sure that the correct replacement parts are obtained for your particular car. Many changes are made during the production run of any car, and when ordering spare parts, it will usually be necessary to know the year the car was built, the model type (eg 'Popular' or 'XR2') and engine size, and in some cases the Vehicle Identification Number (VIN). The VIN is stamped on a metal plate attached to the right-hand side of the engine compartment bulkhead (when viewed from the driver's seat), or on the front crossmember directly behind the right-hand headlight. The VIN usually also appears on the Vehicle Registration Document.

All of the components required for servicing will be available from an official Ford dealer, but most parts should also be available from good motor accessory shops and motor factors, possibly at lower prices.

SERVICING 83

▲ Underbonnet component locations on a 'Mark 2' CVH engine model (air cleaner removed for clarity)

1. Cooling system pressure cap
2. Engine oil level dipstick
3. Carburettor
4. Oil filter (below carburettor)
5. Ignition coil
6. Brake fluid reservoir
7. Battery earth terminal
8. Windscreen washer fluid reservoir
9. Camshaft drivebelt plastic cover
10. Distributor cap and HT leads
11. Number 1 spark plug HT lead
12. Alternator
13. Engine oil filler cap
14. Vehicle Identification Number (VIN) plate
15. Radiator cooling fan shroud
16. Thermostat housing

SERVICE TASKS

The following pages provide easy-to-follow instructions to enable you to carry out the regular service tasks recommended by the manufacturer.

Identifying OHV and CVH engines

To identify a CVH (Compound Valve angle, Hemispherical combustion chambers) engine, look at the right-hand end of the engine (when viewed from the driver's seat). CVH engines have a moulded plastic cover on the end of the engine which covers the camshaft drivebelt. If there's no plastic cover, then the engine is an OHV (OverHead Valve) type.

SERVICING

Every 250 miles (400 km) or weekly

Refer to 'Regular checks' on page 65 and carry out all the checks described.

Every 6000 miles (10 000 km) or 6 months – whichever comes first

ENGINE COMPARTMENT

Check all hoses and pipes, and the surfaces of all components for signs of fluid leakage, corrosion or deterioration

If there's any sign of leaking fluid, try to find the source of the leak and seek advice if necessary. It can be difficult to identify leaking fluids, but if there's obviously a major leak, or if you suspect even a slight brake fluid or petrol leak, don't drive the car until the problem has been investigated by someone suitably qualified.

Check all rubber or fabric hoses for signs of cracking, damage due to rubbing on other components, and general deterioration. It's sensible to have any suspect hoses renewed as a precaution against possible failure.

Check for obvious signs of corrosion on metal pipes, hose clips, and component joints, which may indicate a fluid leak. Any badly corroded components should be cleaned and, if necessary, renewed. Pay particular attention to metal fluid pipes (such as brake pipes), and have them renewed if they're badly pitted or corroded.

ENGINE

Renew the engine oil and filter

Note: The following items will be required for this task:
- A suitable quantity of engine oil of the correct type – refer to **'Service specifications'** on page 61
- New oil filter (of the correct type for your particular car)
- New oil drain plug sealing washer (depending on the condition of the old one)
- Suitable container to catch the old oil as it drains (make sure that the capacity of the container is larger than the oil capacity of the engine – refer to **'Service specifications'** on page 62)
- Oil filter wrench – for removal of old oil filter
- Small quantity of clean rag
- Suitable spanner or socket to fit oil drain plug

The oil should be drained when the engine is warm, immediately after a run. Ideally, the front of the car should be jacked up to improve access. If the front of the car is to be raised, apply the handbrake, then jack up the front of the car and support it securely on axle stands (refer to 'Jacking and vehicle support' in 'Safety first!' on page 80 for details of where to position the jack and axle stands).

Position a suitable container under the engine sump to catch all the oil which will be drained.

Remove the oil filler cap from the top of the engine, then working under the car, unscrew the drain plug from the rear of the engine sump using a suitable spanner or a socket. Oil will be released before the drain plug is removed completely, so take precautions against scalding, as the oil will be hot. Position the container to allow for the fact that the initial jet of oil may fall at least eight inches away from the drain hole, and try not to drop the drain plug when removing it.

▲ *Engine oil drain plug (CVH engine)*

FORD FIESTA

SERVICING 85

Allow the oil to drain for at least 15 minutes, moving the container slightly if necessary from its original position.

Check the condition of the oil drain plug sealing washer, and renew it if necessary.

When the oil has finished draining, clean the drain plug, washer, and the mating face of the sump, then refit the drain plug. Tighten the drain plug securely, but without using excessive force – remember that it could be you who has to undo it next time!

Position a suitable container under the oil filter at the rear of the engine, and unscrew the filter, using a suitable oil filter wrench if necessary to loosen it. Ensure that the oil filter rubber sealing ring comes away with the old filter.

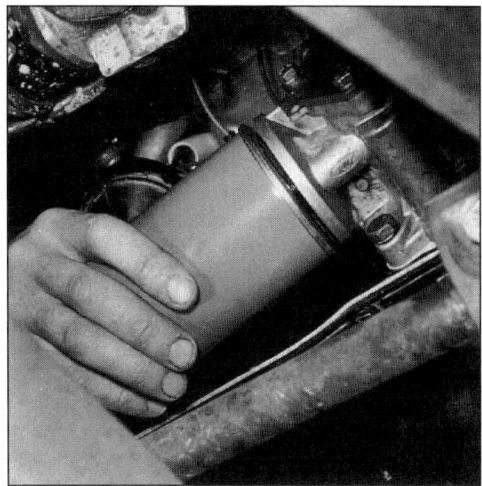

▲ Fitting a new oil filter (OHV engine viewed from underneath the car)

▲ Removing the oil filter using an oil filter removal wrench (CVH engine)

Wipe clean the oil filter mounting flange on the engine. Smear a little clean engine oil on the sealing ring of the new oil filter, then screw the filter onto the engine, and tighten it *by hand only*. If no instructions are provided with the filter, tighten it until the sealing ring touches the mounting face on the engine, then tighten it a further three-quarters of a turn.

Where applicable, lower the car to the ground and, with the car parked on level ground, fill the engine with the correct quantity and grade of oil through the filler on top of the engine. Fill the engine slowly over a period of several minutes, until the level reaches the 'MAX' mark on the dipstick.

Make sure that the oil filler cap is fitted on completion.

Don't race the engine when starting it for the first time after an oil change, as the oil will take a few seconds to circulate. There may be a delay before the oil pressure warning light goes out, as the engine lubrication system fills with oil.

Run the engine and check for leaks from the filter and the drain plug, then stop the engine and check the oil level. Refer to 'Regular checks' on page 66 for details of how to check the oil level. It is likely that a small quantity of oil will have to be added.

Dispose of the old engine oil safely (refer to 'Safety first!' on page 80). **Don't** pour it down a drain.

IGNITION SYSTEM

Clean the distributor cap (where applicable), coil and HT leads

Remove the distributor cap, which may be secured with spring clips or screws.

Thoroughly clean the cap inside and out with a dry lint-free cloth.

Examine the inside of the distributor cap, and if any of the four HT lead contacts inside the cap appear badly burnt or pitted, or if there are any signs of hairline cracks in the plastic, renew the cap.

SERVICING

Make sure that the carbon contact in the centre of the cap is not excessively worn, that it's free to move, and that it protrudes from its holder – if not, renew the cap.

▲ *HT lead contacts inside distributor cap **(A)**, and carbon centre contact **(B)***

Identify the spark plug leads by wrapping sticky labels around them, numbered from 1 to 4, counting from the right-hand end of the engine (when viewed from the driver's seat). The number 1 spark plug is also shown on both the underbonnet component location illustrations earlier in this Section. Trace each plug lead through to the connection on the distributor cap, and attach a numbered sticky label next to the connection on the cap. Similarly identify the coil HT lead, if wished.

Disconnect the HT leads from the coil, distributor cap, and the spark plugs.

Wipe the coil clean and check the contacts on the coil, distributor cap, and the ends of the HT leads for corrosion.

Carefully clean away any corrosion, and wipe the leads clean over their entire length before refitting.

Make sure that the HT leads are refitted in their correct positions, as noted before removal.

Remove, check and, if necessary, adjust the spark plugs

Note: *A suitable spark plug spanner and a spark plug gap adjustment tool, or a set of feeler gauges will be required for this task.*

The spark plugs must be removed with the engine *cold*.

The spark plugs and their HT leads are numbered 1 to 4, starting from the right-hand end of the engine (when viewed from the driver's seat) – the number 1 spark plug is shown on both the underbonnet component location illustrations in this Chapter. The plugs can be checked in any order you like, as long as they're all done; for simplicity, it's probably best to work from one end of the engine to the other.

Starting with the number 1 spark plug, pull the HT lead from the plug – pull on the rubber insulator at the end of the lead, not on the lead itself.

Before removing the spark plug, brush any grit from the area around the plug recess to avoid it dropping into the engine as the plug is removed.

Using a spark plug spanner, unscrew the spark plug and remove it from the engine.

If the correct type of spark plugs is used for the engine, and the engine is in good condition, the spark plugs should not need attention between scheduled renewal intervals (in this case, every 12 000 miles (20 000 km) or 12 months).

If the spark plug electrodes are heavily fouled (very dirty), it may be advisable to renew all four plugs, especially if poor starting has been experienced – refer to the *'12 000 miles (20 000 km)'* service Section on page 90. Spark plug cleaning should only be carried out by someone suitably qualified using specialised equipment, otherwise the plugs may easily be damaged.

If the plug appears to be in good condition, check the electrode gap as follows before refitting it (refer to *'Service specifications'* on page 63 for the correct electrode gap). The size of the spark plug electrode gap is extremely important, and it can drastically affect the performance of the engine.

Measure the gap between the electrodes using a spark plug gap adjustment tool, or a feeler gauge.

If adjustment is required, carefully bend the **outer** electrode until the correct gap is obtained. **Never** try to bend the centre electrode.

Make sure that the plug threads and the

FORD FIESTA

SERVICING 87

▲ Checking a spark plug gap using a suitable tool

▲ Adjusting a spark plug gap using the correct tool

seating areas in the cylinder head are clean, and apply a little light oil to the plug threads. Screw the plug in by hand initially.

Using the spark plug spanner, tighten the plug until initial resistance is felt, then tighten by a further one-sixteenth of a turn for the taper-seat type plugs fitted to OHV engines, or one quarter of a turn for the washer type plugs fitted to CVH engines. **Do not** overtighten the plug.

Refit the HT lead, making sure that it is a secure fit over the end of the plug – there should be a soft 'click' as the lead is connected.

Repeat the above procedures for the remaining three spark plugs in turn, removing only one HT lead at a time, to avoid confusion.

On completion, check once more that all four HT leads are securely fitted.

On models with a contact breaker points distributor, lubricate the distributor

Remove the distributor cap, and wipe clean the cam on the distributor shaft which operates the contact breaker points.

Apply a trace of high melting point grease to each of the four cam lobes **(2)**. The grease is best applied using the tip of a screwdriver or a similar tool.

Apply two drops (no more) of light oil to the felt pad at the top of the distributor shaft.

▲ Applying grease **(1)** to a distributor cam lobe **(2)**

▲ Using an oil can **(A)** to apply oil to the distributor shaft felt pad **(B)**

FORD FIESTA

SERVICING

On models with a contact breaker points distributor, check and, if necessary, adjust the points gap

Note: *A set of feeler gauges will be required for this check.*

Remove the distributor cap, then pull the rotor arm from the distributor shaft.

Using a screwdriver, gently prise the contact breaker points open to examine the condition of the contact faces.

If the contacts are rough, pitted, or excessively dirty, the points should be renewed as described in the '*12 000 miles (20 000 km)*' service Section on page 90.

Assuming that the points are in good condition, the gap between the two contact faces should be checked and if necessary adjusted. This can be done using feeler gauges as follows, although it's preferable to use the more accurate 'dwell angle' method described in the Owners Workshop Manual.

To check the points gap using feeler gauges, turn the crankshaft until the heel of the contact breaker arm is resting on the peak of one of the four lobes (or 'corners') on the distributor shaft cam, and the points are fully open. The crankshaft can be turned by engaging 3rd gear and pushing the car (with the handbrake released).

Select a feeler gauge of thickness equal to the specified points gap (refer to '*Service specifications*' on page 63), and slide it

▲ *Using a feeler gauge to measure the contact breaker points gap (Bosch distributor)*

between the two contact faces. The feeler gauge should be a firm sliding fit.

If adjustment is required, slacken the points retaining screw(s) slightly (one screw on Bosch distributors, two screws on Ford distributors), so that the fixed point can just be moved to give the required gap.

The easiest way of moving the fixed point is to engage a screwdriver in the slot on the end of the fixed point, and lever against the corresponding slot or pips on the distributor baseplate.

After adjustment, tighten the retaining screw and recheck the gap.

Refit the rotor arm and the distributor cap.

▲ *Using a feeler gauge to measure the contact breaker points gap (Ford distributor)*

▲ *Contact breaker points gap **(A)** – Ford distributor*

FORD FIESTA

SERVICING 89

▲ Contact breaker points gap **(A)** – Bosch distributor

BRAKING SYSTEM

Check the operation of the low brake fluid level warning system

Open the bonnet, then unscrew the filler cap from the top of the brake hydraulic fluid reservoir, and allow the float under the filler cap to hang downwards (if necessary disconnect the wiring from the cap to enable the cap to be removed, but reconnect the wiring once the cap has been removed). Hold a piece of rag or a cloth under the filler cap to collect any drips of hydraulic fluid.

On models which have a combined handbrake 'on'/low brake fluid level warning light, chock the wheels to prevent the car from moving, then switch on the ignition and release the handbrake.

The warning light should illuminate and, assuming that the fluid level is not low, should then go out when the filler cap is refitted to the fluid reservoir (with the handbrake still released where applicable).

If the warning light doesn't operate as described, it's likely that the circuit is faulty, and the problem should be investigated by someone suitably qualified.

Where applicable, re-apply the handbrake and remove the chocks from the wheels on completion.

Refit the filler cap using a reversal of the removal procedure.

WHEELS

Check the tightness of the roadwheel bolts

Using the wheel brace (refer to 'Changing a wheel' on page 38), check that the bolts cannot be turned easily using reasonable force. Do not overtighten the bolts.

INTERIOR AND BODYWORK

Check the seat belt webbing for cuts and damage, and check the seat belt operation

If any of the seat belt webbing is damaged, frayed or worn, the relevant seat belt(s) must be renewed.

Check that all inertia reel seat belts can be pulled smoothly from their retractors, and that they return smoothly, unassisted.

Check that the locking mechanism works correctly by sharply tugging the seat belt webbing.

If a seat belt retractor or locking mechanism appears to be faulty, the relevant seat belt should be renewed.

Lubricate all hinges, door locks, and the bonnet release mechanism

Use general-purpose light oil for the door locks, and a suitable general-purpose grease for the hinges and bonnet release mechanism.

Don't use an excess of lubricant, or it may find its way onto other components, or onto the clothing of the car's occupants!

▲ Removing the filler cap from the brake fluid reservoir

SERVICING

Carefully inspect the paintwork for damage, and the bodywork for corrosion

Touch up any minor stone chips, and have any corroded areas repaired before the damage becomes serious.

For details of how to treat minor paint scratches, refer to *'Bodywork and interior care'* on page 106.

CAR UNDERSIDE

Check all hoses and pipes, and the surfaces of all components for signs of fluid leakage, corrosion, damage or deterioration

Refer to *'ENGINE COMPARTMENT'* at the beginning of the *'6000 miles (10 000 km)'* service Section on page 84.

In addition, check the exhaust system and the fuel tank for any sign of damage or serious corrosion (light surface rust is to be expected).

Check the exhaust mountings to make sure that they're secure.

Check the visible suspension and steering components for obvious signs of damage or wear.

Check the driveshafts for obvious signs of distortion or damage.

Pay particular attention to the driveshaft and steering gear rubber gaiters, which should be renewed if they're split, or if there are any obvious signs of lubricant leakage.

▲ *Check the driveshaft and steering gear rubber gaiters for splits (arrowed)*

ADDITIONAL TASKS

Note that as well as the tasks described in the preceding paragraphs, the additional tasks given in the *'Service schedule'* chart on page 74 should be carried out every 6000 miles (10 000 km) or 6 months – whichever comes first. These tasks require more detailed explanation, or the use of special tools, and are considered beyond the scope of this Handbook. For details of these additional tasks, refer to the Owners Workshop Manual.

Every 12 000 miles (20 000 km) or 12 months – whichever comes first

In addition to all the items in the *'6000 miles (10 000 km)'* service on page 84, carry out the following.

IGNITION SYSTEM

Renew the spark plugs

Note: *Ensure that the correct type of new spark plugs is obtained for your particular car.*

Follow the procedure given under *'Remove, check and, if necessary, adjust the spark plugs'* in the *'6000 miles (10 000 km)'* service on page 86, but fit new spark plugs.

Make sure that the electrode gap is set correctly – refer to *'Service specifications'* on page 63 for the correct gap.

On contact breaker points distributors, renew the contact breaker points

Note: *Ensure that the correct type of new contact breaker points are obtained for your particular car.*

Remove the distributor cap, then pull the rotor arm from the distributor shaft.

Disconnect the contact breaker points wire. On Bosch type distributors, the wire can simply be pulled from the spade connector. On Ford type distributors, the wire securing screw must be loosened before the forked connector can be freed.

Unscrew the contact breaker points retaining screw(s) (one screw on Bosch distributors, two screws on Ford distributors), and withdraw the points from the distributor. Take care not to drop the screw(s) and washer(s) into the distributor. If possible use a magnetic screwdriver, or hold the screw on the end of the screwdriver using a blob of grease.

▲ *Ford distributor*
  A Contact breaker points wire securing screw
  B Contact breaker points retaining screws

▲ *Bosch distributor*
  A Contact breaker points wire
  B Contact breaker points retaining screw

Wipe clean the cam on the distributor shaft which operates the contact breaker points, then apply a trace of high-melting-point grease to each of the four cam lobes. The cam lobes are the four raised 'corners' on the distributor shaft cam.

Apply two drops (no more) of light oil to the felt pad at the top of the distributor shaft.

New contact breaker points are usually supplied coated with a protective film. This should be wiped from the points before fitting, using a cloth moistened with methylated spirit.

Locate the new contact breaker points in the distributor, and lightly tighten the retaining screw(s).

Reconnect the wire, then adjust the points gap as described in the '*6000 miles (10 000 km)*' service on page 88.

MANUAL GEARBOX

Check and if necessary top-up the gearbox oil level

Note: *A suitable spanner or socket, or on later models a Torx or Allen key (depending on model), will be required to remove and refit the oil filler plug*

With the car on level ground, wipe around the filler plug which is located at the front of the gearbox casing. Unscrew the plug using a suitable spanner, or a Torx or Allen key, as applicable, on later models. The plug can be reached from above or below the car.

Locate the aluminium build code tag, which is secured to one of the upper gearbox housing bolts, and note the gearbox part number stamped on the tag.

If the last letter of the part number suffix is a 'D' (gearboxes built up to August 1985), the oil level must be maintained between 5 and 10 mm (0.2 and 0.4 in) below the lower edge of the filler plug hole.

If the last letter of the part number suffix is an 'E' (gearboxes built after August 1985), the oil level must be maintained between 0 and 5 mm (0 and 0.2 in) below the lower edge of the filler plug hole.

To make level checking easier, a dipstick can be made from a thin rod bent at right-angles and having marks on one 'leg' made with a file at 5 mm (0.2 in) intervals. Be sure to make the dipstick large enough not to fall into the gearbox, and of a material strong enough that the end of the dipstick doesn't break off inside the gearbox.

Insert the dipstick so that the unmarked leg rests on the lower edge of the filler plug hole, with the marked end immersed in the oil.

SERVICING

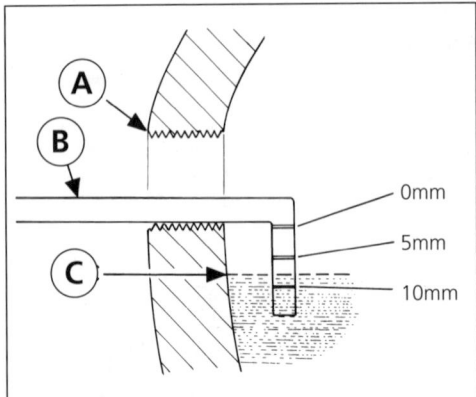

▲ Checking the manual gearbox oil level using a home-made dipstick
A Gearbox filler plug hole
B Dipstick
C Oil level (typical level shown – see text)

Remove the dipstick, read off the oil level and top-up if necessary using the specified grade of oil (refer to *'Service specifications'* on page 61).

Refit and tighten the filler plug on completion.

▲ Topping-up the manual gearbox oil level (viewed from underneath the car)

AUTOMATIC TRANSMISSION

Check and if necessary top-up the automatic transmission fluid level

The fluid level must be checked with the engine and transmission at normal operating temperature, preferably after a short journey.

Park the car on level ground, then apply the handbrake.

With the engine running at its normal idle speed, apply the brakes, while moving the transmission selector lever through the full range of positions three times, then back to the **P** position.

Allow the engine to idle for a further minute.

With the engine still idling, pull out the automatic transmission fluid level dipstick (located at the front left-hand side of the engine compartment when viewed from the driver's seat) and wipe it dry with a clean lint-free rag.

Fully insert the dipstick, then withdraw it again and check the fluid level, which should be between the 'MAX' and 'MIN' marks.

▲ Automatic transmission fluid level dipstick markings

If topping-up is necessary, use only the specified fluid type (refer to *'Service specifications'* on page 61).

Top-up by pouring the fluid through the dipstick tube. Take care not to overfill; the level **must not** exceed the 'MAX' mark.

If there's a need for regular topping-up, it's likely that there's a fluid leak, and the problem should be referred to a Ford dealer.

BRAKING SYSTEM

Check the operation of the handbrake mechanism, and lubricate the components

Engage 1st gear (manual gearbox models) or **P** (automatic transmission models), as applicable, then chock the front wheels.

Jack up the rear of the car, and support it securely on axle stands (refer to *'Safety first!'*

FORD FIESTA

SERVICING 93

▲ Handbrake components ('Mark 1' models)

A Handbrake lever
B Front cable
C Cable adjuster
D Rear cable
E Equaliser
F Rear cable

on page 80 for details of the correct jacking procedure).

Apply the handbrake and check that both rear wheels are locked, then release the handbrake and check that both rear wheels can be turned by hand (there may be some resistance, but it should be possible to turn the wheels without excessive force).

If the wheels don't lock with the handbrake applied, or cannot be turned with the handbrake released, the handbrake mechanism is faulty, possibly due to lack of lubrication or dirty components.

Examine the exposed handbrake cable components. The cables run from beneath the handbrake lever to the equaliser bracket under

▲ Lubricate both sides of the rear handbrake cable guide (arrowed) – 'Mark 1' model

▲ Lubricate both sides of the handbrake cable guides (arrowed) – 'Mark 2' model

FORD FIESTA

94 SERVICING

▲ *Handbrake components ('Mark 2' models)*
  A *Cable guide*   **B** *Cable adjuster*   **C** *Rear brake shoe operating assembly*

the car, from where the cable routing is directed to each side of the car to operate the two rear brake assemblies.

Clean away any road dirt and corrosion, and examine all the components for obvious signs of wear or damage. If any of the components are excessively worn or damaged, seek advice and if necessary have them renewed.

Apply a small amount of grease to the exposed moving parts of the handbrake linkage. Pay particular attention to the equaliser components, and the cable guides in which the cables slide.

Check that all the components move freely and operate correctly, and wipe away any excess grease.

On completion, lower the car to the ground and remove the chocks from the front wheels.

ELECTRICAL SYSTEM

Check the battery lead connections for security

Disconnect the battery leads (always disconnect the earth (negative) lead first, and reconnect it last), and clean the battery terminals and the lead ends.

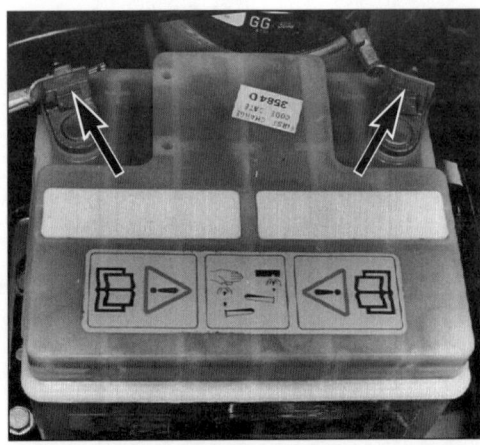

▲ *Make sure that the battery lead connections are secure and clean*

Remove any corrosion using a stiff brush, taking care to avoid debris entering your eyes or contacting your skin (wash your hands thoroughly afterwards).

Refit the leads, making sure that they are secure, and smear the exposed metal surfaces with petroleum jelly (do not use ordinary

FORD FIESTA

grease), which will help to prevent corrosion.

Inspect the battery tray for signs of white-coloured salts – if evident, these should be removed with a brush (wear eye protection) or with hot water (take care to avoid splashes), as they are highly corrosive.

Check the condition and adjustment of the alternator drivebelt

Note: *If the drivebelt requires adjustment, a suitable spanner or socket will be required to fit the alternator mounting and adjuster link bolts.*

Examine the drivebelt for signs of cracking, obvious wear, or contamination, and renew it if necessary (note that on OHV engines, the belt also drives the engine coolant pump).

To check the tension of the belt, press down on the belt midway between the alternator and

▲ *Alternator drivebelt tension checking point (OHV engine)*

▲ *Alternator drivebelt tension checking point (CVH engine)*

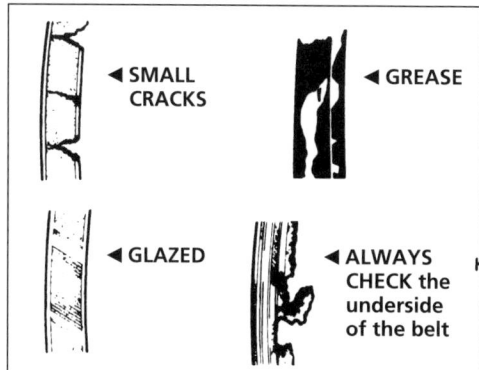
▲ *Examine the alternator drivebelt for signs of wear, deterioration or contamination*

▲ *Alternator mounting and adjuster link bolts (OHV engine)*

water pump pulleys on OHV engines, or midway between the alternator and crankshaft pulleys on CVH engines.

The deflection of the belt should be approximately 10 mm (0.4 in) under moderate finger pressure.

To adjust the tension, slacken the alternator mounting bolts and the adjuster link bolts, and move the alternator towards or away from the engine as necessary to give the correct tension.

Once the tension is correct, tighten the adjuster link-to-alternator bolt, adjuster link-to-engine bolt, and the alternator front and rear mounting bolts in that order.

To renew the belt, slacken all four mounting and adjuster bolts, and move the alternator towards the engine to remove the belt.

Fit the new belt, then pull the alternator away from the engine until the belt is fairly tight, and tighten the adjuster link-to-alternator bolt.

Tension the belt as described previously.

If a new belt has been fitted, the tension should be rechecked after the engine has run for approximately ten minutes.

SERVICING

ROAD TEST

Check the operation of all instruments and electrical equipment

Make sure that all instruments read correctly, and switch on all electrical equipment in turn to check that it functions properly.

Check for any abnormalities in the steering, suspension, handling or road feel

Drive the car and check that there are no unusual vibrations or noises.

Check that the steering feels positive, with no excessive 'sloppiness', or roughness, and check for any suspension noises when cornering and driving over bumps.

Check the performance of the engine, clutch (where applicable), gearbox and driveshafts

Listen for any unusual noises from the engine, clutch and gearbox.

Make sure that the engine runs smoothly when idling, and that there's no hesitation when accelerating.

Check that, where applicable, the clutch action is smooth and progressive, that the drive is taken up smoothly, and that the pedal travel is not excessive. Also listen for any noises when the clutch pedal is depressed.

On manual gearbox models, check that all gears can be engaged smoothly without noise, and that the gear lever action is not abnormally vague or 'notchy'.

On automatic transmission models, check that all the gear positions can be selected with the car at rest. If any problems are found, they should be referred to a Ford dealer.

Listen for a metallic clicking sound from the front of the car as the car is driven slowly in a circle with the steering on full lock. Carry out this check in both directions. If a clicking noise is heard, this indicates wear in a driveshaft joint, in which case seek advice and renew the joint if necessary.

Check the operation and performance of the braking system

Make sure that the car does not pull to one side when braking, and that the wheels do not lock prematurely when braking hard.

Check that there's no vibration through the steering when braking.

Check that the handbrake operates correctly without excessive movement of the lever, and that it holds the car stationary on a slope.

ADDITIONAL TASKS

Note that as well as the tasks described in the preceding paragraphs, the additional tasks given in the *'Service schedule'* chart on page 74 should be carried out every 12 000 miles (20 000 km) or 12 months – whichever comes first. These tasks require more detailed explanation, or the use of special tools, and are considered beyond the scope of this Handbook. For details of these additional tasks, refer to the Owners Workshop Manual.

Every 24 000 miles (40 000 km) or 1 year – whichever comes first

---

AUTOMATIC TRANSMISSION

Renew the automatic transmission fluid

Refer to a Ford dealer.

Every 24 000 miles (40 000 km) or 2 years – whichever comes first

---

In addition to the items in the *'6000 miles (10 000 km)* service on page 84' and *'12 000 miles (20 000 km)'* service on page 90, carry out the following.

ENGINE

On OHV engine models, clean the crankcase emission control filter

'Mark 1' models

The filter is an integral part of the engine oil filler cap.

To clean the filter, disconnect the hoses from the oil filler cap, noting their locations, then remove the filler cap.

FORD FIESTA

SERVICING

The filler cap should be washed in paraffin to remove any sediment build-up in the cap itself or the filter.

Thoroughly dry the filler cap before refitting it and reconnecting the hoses. Make sure that the hoses are refitted to their correct locations as noted before removal.

'Mark 2' models

The filter is located in the engine oil filler cap.

To clean the filter, disconnect the oil filler cap-to-inlet manifold hose, and remove the filter from the filler cap by pulling and twisting it free.

▲ *Crankcase emission control filter renewal (CVH engine model)*

FUEL AND EXHAUST SYSTEMS

Renew the air cleaner filter element

The air cleaner is located on the top of the carburettor at the rear of the engine.

Unscrew and remove the air cleaner cover retaining screw(s) or nuts as applicable.

Where applicable, release the cover retaining clips around the side of the air cleaner body, then remove the cover and lift out the air cleaner element.

Wipe out the inside of the air cleaner body and the cover, then place the new element in position. Refit the cover using a reversal of the removal procedure.

▲ *Removing the crankcase emission control filter ('Mark 2' model with OHV engine)*

Wash the filter in paraffin to remove any sediment build-up.

Thoroughly dry the filter before refitting it to the filler cap and reconnecting the hose.

If the filter is severely contaminated or damaged, it should be renewed.

On CVH engine models, renew the crankcase emission control filter

Where fitted, the filter is located in the underside of the air cleaner body.

To renew the filter, disconnect the hoses from the filter, noting their locations, then pull the filter from the air cleaner.

Make sure that the sealing grommet is in position in the air cleaner body, before pushing the new filter into position and reconnecting the hoses in their correct positions.

▲ *Removing the air cleaner cover ('Mark 1' models)*
  A Air cleaner cover
  B Filter element

FORD FIESTA

98 SERVICING

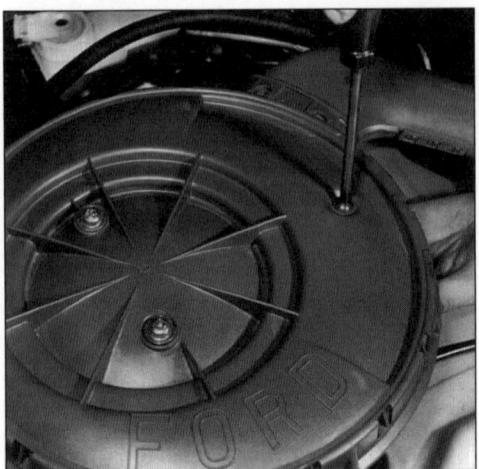

▲ *Removing an air cleaner cover retaining screw ('Mark 2' model)*

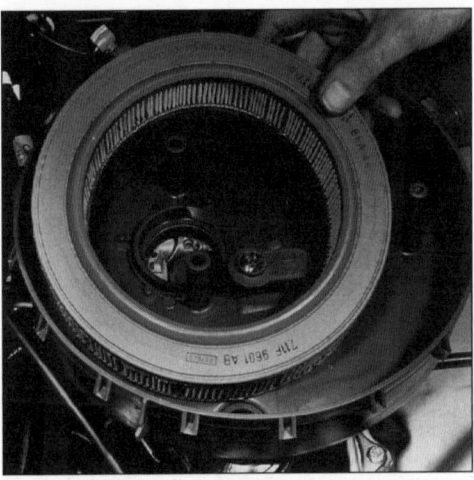

▲ *Lifting out the air cleaner element ('Mark 2' model)*

ADDITIONAL TASKS

Note that as well as the tasks described in the preceding paragraphs, the additional tasks given in the *'Service schedule'* chart on page 74 should be carried out every 24 000 miles (40 000 km) or 2 years – whichever comes first. These tasks require more detailed explanation, or the use of special tools, and are considered beyond the scope of this Handbook. For details of these additional tasks, refer to the Owners Workshop Manual.

Every 36 000 miles (60 000 km)

In addition to the items listed in the previous services, the additional engine task given at the end of the *'Service schedule'* chart on page 76 should be carried out every 36 000 miles (60 000 km). For details, refer to the Owners Workshop Manual.

Every 36 000 miles (60 000 km) or 2 years – whichever comes first

COOLING SYSTEM
Renew the coolant
Note: *The following items will be required for this task:*
- A suitable quantity of coolant solution (made up from approximately 55% clean water and 45% antifreeze)
- Suitable container to catch the old coolant as it drains
- Suitable spanner or socket to fit the cylinder block drain plug (where applicable)
- A suitable screwdriver to fit the radiator drain plug or radiator hose clip(s), as applicable

Draining the system
The coolant should be drained with the engine cold.

Set the heater temperature control inside the car to the maximum-heat position.

Remove the cooling system pressure cap. Refer to the *'Checking coolant level'* procedure on page 67 for details of the cap location and type.

Check to see if a drain plug is fitted to either lower corner of the radiator.

If so, position a suitable container underneath the radiator, making sure that the container is big enough to hold all the coolant, then unscrew the drain plug and allow the coolant to drain.

If no drain plug is fitted, position the container beneath the radiator bottom hose connection, then slacken the hose clip, release the hose, and allow the coolant to drain.

FORD FIESTA

SERVICING

▲ *Radiator drain plug (arrowed) – 'Mark 2' OHV engine model*

A cylinder block drain plug is also fitted to certain models, on the forward facing side of the engine, towards the gearbox end. Where a drain plug is fitted, reposition the container, then unscrew the plug and allow the remaining coolant in the cylinder block to drain into the container.

▲ *Cylinder block drain plug (arrowed) – OHV engine model*

If rust or sludge is evident in the coolant which has been drained, the cooling system should be flushed to remove any further contamination from the engine and radiator, as described below.

If the coolant drained is clean, then the cooling system can be refilled as described later in this Section.

Flushing the system
Note: *A new thermostat housing gasket will be required for this task, and on CVH engine models a new thermostat sealing ring will be required*

To flush the cooling system, first the thermostat must be removed.

On OHV engines the thermostat is located in a housing at the right-hand end of the engine, and on CVH engines the thermostat is located in a housing at the left-hand end of the engine (when viewed from the driver's seat).

Slacken the clips and disconnect the coolant hoses from the thermostat housing.

Disconnect the wiring from the radiator fan switch, which is located in the thermostat housing.

Unscrew the securing bolts (two on OHV engines and three on CVH engines), and remove the thermostat housing cover. If it's stuck, tap it gently with a soft-faced mallet or similar tool (Don't use a metal-faced hammer).

▲ *Thermostat housing cover securing bolts (arrowed) – CVH engine*

▲ *Removing the thermostat housing . . .*

FORD FIESTA

SERVICING

▲ . . . and the thermostat (OHV engine model)

On OHV engines, the thermostat can now be lifted from its location in the cylinder block. If the thermostat is stuck in position, don't lever it out, but cut around the edge with a sharp knife.

On CVH engines, the thermostat is retained in the thermostat housing cover, and there is no need to remove it once the cover has been removed from the engine.

Insert the end of a garden hose into the thermostat housing in the cylinder block, and flush the system through using cold water.

Continue flushing until clean water flows from the disconnected radiator bottom hose or radiator drain plug, and the cylinder block drain plug, as applicable.

If it seems that the system is particularly badly contaminated, and flushing does not appear to be solving the problem, seek advice, as it may be possible to clean the radiator using a proprietary cleaning compound.

When flushing is complete, refit the thermostat.

Carefully remove all traces of old gasket from the faces of the thermostat housing and cover, then refit the cover using a new gasket, and tighten the securing bolts.

Reconnect the hoses and the wiring, and tighten the hose clips.

Refilling the system

Reconnect the radiator bottom hose and tighten the clip, or refit and tighten the radiator drain plug, and the cylinder block drain plug, as applicable.

A solution of 55% clean water and 45% antifreeze should be used to fill the system all year round. It's important to note that because of the different types of metals used in the engine, it's vital to use antifreeze with suitable anti-corrosion additives all year round. **Never** use plain water to fill the whole system. If there's any doubt about the strength of the coolant mixture, it's better to add too much antifreeze than not enough.

▲ Filling the cooling system through the radiator – 'Mark 1' models

On 'Mark 1' models, the system must be vented whilst refilling to prevent the build-up of airlocks. To vent the system, identify the coolant hose which runs from the heater connection on the engine compartment rear bulkhead to the connection on the inlet manifold (almost directly below the carburettor). This hose should be disconnected at the inlet manifold. Fill the system through the radiator filler neck, whilst holding the disconnected heater coolant hose at the same height as its connection on the inlet manifold. When a steady stream of coolant, free of air bubbles, flows from the disconnected hose, quickly reconnect the hose to the inlet manifold. Continue to fill the system through the radiator, until the coolant level is within 30 mm (1.2 in) of the flange of the radiator filler neck, then refit the pressure cap to the radiator, turning it firmly clockwise to lock it into position. Check the level of the coolant in the expansion tank, and top-up to the 'MAX' mark if necessary.

On 'Mark 2' models, fill the system slowly through the expansion tank until the level reaches the 'MAX' mark. Wait a few moments

SERVICING 101

▲ Filling the cooling system through the expansion tank – 'Mark 2' models

for trapped air to escape, then add more coolant. Repeat this procedure until the level does not drop any more, then refit the pressure cap.

On all models, start the engine and run it to normal operating temperature, then check that the heater works, and switch off the engine.

Once the engine has cooled, squeeze the hoses and check for air locks (it should be possible to feel warm coolant in all the hoses).

If the heater failed to give warm air, there may be an air lock in the heater, in which case the air can be forced out by squeezing the heater hoses (which run to the bulkhead at the back of the engine compartment).

If necessary, air can be removed from a hose by loosening a hose clip, and squeezing the hose until all the air is forced out (obviously some coolant will be forced out too).

Recheck the coolant level and carry out any final topping-up as described in *'Checking coolant level'* on page 67.

Every 36 000 miles (60 000 km) or 3 years – whichever comes first

In addition to the items listed in the previous services, the additional braking system and steering tasks given in the *'Service schedule'* chart on pages 74 & 76 should be carried out every 36 000 miles (60 000 km) or 3 years – whichever comes first. For details, refer to the Owners Workshop Manual.

SEASONAL SERVICING

If you carry out all the procedures described in the previous Section, at the recommended mileage or time intervals, then you'll have gone a long way towards getting the best out of your car in terms of both performance and long life. In spite of this, there are always other areas, not dealt with in regular servicing, where neglect can spell trouble.

A little extra time spent on your car at the beginning and end of every Winter will be well worthwhile in terms of peace of mind and prevention of trouble. The suggested tasks which follow have therefore been divided into Spring and Autumn Sections – but it's always a good idea to do them more frequently if you feel able.

Autumn

COOLING SYSTEM

Check the radiator and all hoses for signs of deterioration or damage

Refer to *'ENGINE COMPARTMENT'* in the *'6000 miles (10 000 km)'* service Section on page 84.

Check the strength of the coolant solution

This can be done using a coolant hydrometer, which should be available from most motor accessory shops and motor factors.

The hydrometer will give an indication of the amount of antifreeze in the system, and should

SERVICING

be used exactly in accordance with the hydrometer manufacturer's instructions.

The strength of the solution should never be allowed to drop below 55% clean water and 45% antifreeze.

If there is any doubt about the strength of the coolant solution, the system should be drained and refilled as in the '36 000 miles (60 000 km) or 2 years' service Section on page 98.

FUEL SYSTEM

On 'Mark 1' models, reset the position of the air cleaner intake spout

On some models, the air cleaner intake spout has two positions for Summer and Winter conditions. This type of air cleaner can be recognised by the clip assembly on the spout, as shown in the accompanying illustration.

When the air temperature is low, during late Autumn and Winter, the intake spout should be set so that it draws in warm air from the area around the engine exhaust manifold.

When the air temperature is warmer during late Spring and Summer, the intake spout should be repositioned so that it draws cooler air from the engine compartment.

To change the position of the intake spout, press down the clip **(1)** and turn the intake spout to position **(2)** for Winter operation, or position **(3)** for Summer operation.

▲ Air cleaner intake spout positions ('Mark 1' models) For **1**, **2** and **3** see text

ELECTRICAL SYSTEM

Where applicable, check the battery electrolyte level

Refer to 'Regular checks' on page 70.

Check and if necessary clean the battery terminals

Make sure that the terminals are secure, and clean any corrosion from the metal.

To prevent corrosion, the terminals can be coated with petroleum jelly (don't use ordinary grease).

If any white-coloured salts are present on or around the battery, refer to 'ELECTRICAL SYSTEM' in the '12 000 miles (20 000 km)' service Section on page 94.

Check the condition and adjustment of the alternator drivebelt

Refer to the procedure given in the '12 000 miles (20 000 km)' service Section on page 95.

Check the operation of all lights, electrical equipment and accessories

Refer to 'Fault finding' on page 121 if any problems are discovered.

If any of the bulbs need to be renewed, refer to 'Bulb, fuse and relay renewal' on page 107.

Check the washer fluid level and the wipers and washers

Refer to 'Regular checks' on page 70.

TYRES

Check tread depth and condition

Refer to 'Regular checks' on page 68. Remember that you may be driving in slippery conditions during the Winter.

BODYWORK

Thoroughly clean the car

Wash the car, and then polish it thoroughly to help protect the paint during the Winter.

SERVICING 103

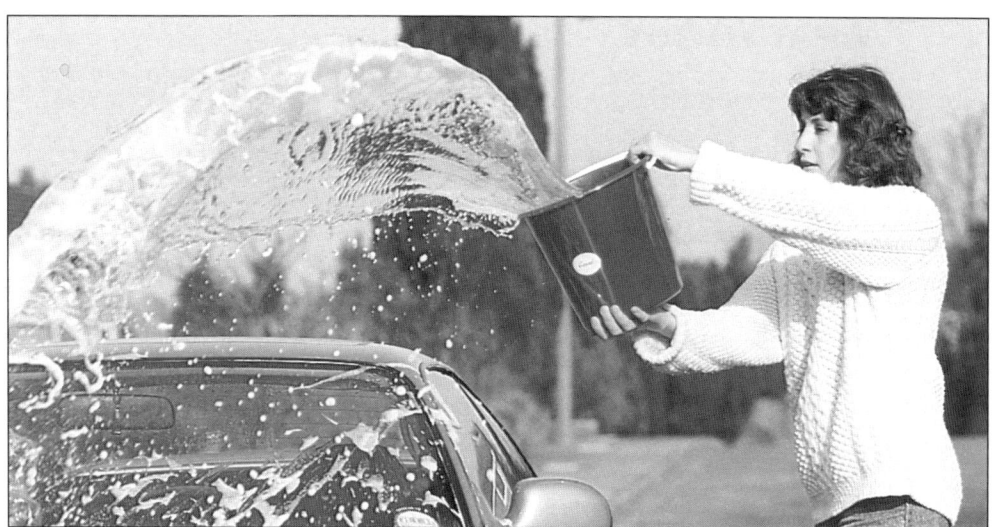

Spring

FUEL SYSTEM

On 'Mark 1' models, reset the position of the air cleaner intake spout

Refer to the procedure given under *'Autumn'* servicing on page 102.

UNDERBODY

Thoroughly clean the underside of the car

The best time to clean the underside of the car is after the car has been driven in wet conditions, when the accumulated dirt will be softened up.

To clean the car, first of all the car must be jacked up as high as possible (making sure that it is safely supported – refer to *'Safety first!'* on page 80).

Gather together a quantity of paraffin, or water-soluble solvent, a stiff-bristle brush, a scraper, and a garden hose.

With the car jacked up and safely supported, get underneath the car, and cover the brake components at each wheel with polythene bags to stop dirt and water getting into them.

Loosen any encrusted dirt, scraping or brushing it away – the paraffin or solvent can be used where there's oil contamination. Pay particular attention to the wheel arches. Take care not to remove the underseal (a waxy substance applied to the car underbody to protect against corrosion).

When all the dirt is loosened, a wash down with the hose will remove the remaining dirt and mud.

You can now check for signs of damage to the underseal. If there's any sign of the underseal breaking away, patch it up by spraying or brushing on a suitable underseal wax (available from motor accessory shops and motor factors). Make sure the area is clean and dry before applying the wax.

Take the opportunity to check for signs of rusting. Likely places are the body sills and floor panels.

If rust is found, seek advice and have the affected area repaired before the problem gets too bad.

On completion, lower the car to the ground.

BODYWORK

Thoroughly check and clean the surfaces of the bodywork

Give the car a thorough wash, and check for stone chips and rust spots. For details of how to treat these minor paint problems, refer to *'Bodywork and interior care'* on page 106.

After any repairs to the paint have been carried out (not before, otherwise the paint will not stick), give the car a polish.

WHAT TO BUY

If you're intending to carry out your own servicing, you'll need to obtain a few basic tools. Although at first sight you may think that tools seem expensive, once you've bought them they should last a lifetime if you look after them properly.

The tools supplied with the car will enable you to change a wheel, and not much more. The absolute minimum tool kit you'll need to carry out any maintenance or servicing will be a range of metric spanners, two screwdrivers (one for crosshead or 'Phillips' type screws), and a pair of pliers. With a bit of ingenuity, these items should enable you to complete the more basic routine servicing jobs, but they won't allow you to do much else.

When buying tools, it's important to bear in mind the quality. You don't need to buy the most expensive tools available, but generally you get what you pay for. Cheap tools may prove to be a false economy, as they're unlikely to last as long as better quality alternatives.

It's very difficult to lay down hard and fast rules on exactly what you're going to need, but the following list should be helpful in building up a good tool kit. Combination spanners (ring one end, open-ended the other) are recommended because, although more expensive than double open-ended ones, they give the advantages of both types.

- Combination spanners to cover a reasonable range of sizes (say 7 mm to 17 mm at least)
- Adjustable spanner
- Spark plug spanner (with rubber insert)
- Spark plug gap adjustment tool
- Set of feeler gauges
- Screwdriver (Plain) – 100 mm long blade x 6 mm diameter (approx)
- Screwdriver (Crosshead) – 100 mm long blade x 6 mm diameter (approx)
- Oil filter wrench
- Pliers
- Tyre pump
- Tyre pressure gauge
- A suitable key (Torx or Allen type, depending on model) to fit the gearbox oil filler plug (later models only)
- Oil can
- Funnel (medium size)
- Stiff brush (for general cleaning jobs)
- Tool box
- Hydraulic jack (ideally a trolley jack)
- Pair of axle stands
- Suitable containers for draining engine oil and coolant
- Inspection lamp

This is by no means a comprehensive list of tools, and you'll probably want to gradually add to your tool box as you discover the need for other tools, especially if you decide to tackle some of the more advanced servicing jobs described in the **Owners Workshop Manual** for your particular car.

CARE OF YOUR TOOLS

Having bought a reasonable set of tools, it's worth taking the trouble to look after them. After use, always wipe off any dirt or grease using a clean, dry cloth before putting them away. Never leave tools lying around after they've been used – a tool rack, or better still a proper toolbox will prove the best way of keeping everything up together. Rags can be wrapped around loose tools to prevent them from rattling if you're going to keep them in the boot of the car.

Feeler gauges should be wiped with an oily cloth from time to time, to leave a thin coating of oil on the metal which will prevent corrosion. Screwdriver blades inevitably lose their keen edges, and a little occasional attention with a file or an oilstone will keep them in good condition.

BODYWORK AND INTERIOR CARE

It's well worth spending the time and effort necessary to look after your car's bodywork and interior. Cleaning your car regularly will not only improve its appearance, it will also protect the bodywork against the elements and the grime encountered in everyday driving. It's worth bearing in mind that if your car looks clean and tidy, it will also be worth more money when you come to sell it or trade it in for a newer model.

This Section will help you to keep your car in 'showroom' condition. If you wish to tackle repairs to more serious bodywork damage or corrosion, comprehensive details can be found in our **'Car Bodywork Repair Manual'**.

CLEANING THE INTERIOR

It's a good idea to clean the interior of the car first, before cleaning the exterior, as this will avoid spreading the dirt from inside over the bodywork.

Start by removing all the loose odds and ends from inside the car, not forgetting the ashtrays, glovebox and any interior pockets. Take out any loose mats or carpets, which should be shaken and brushed, and if possible vacuum-cleaned.

The inside of the car can be cleaned with a brush and dustpan, or preferably a vacuum-cleaner. If you can't use your household vacuum-cleaner, it may be worth investing in one of the small 12-volt hand vacuum-cleaners which can be powered from the car battery. If the carpets are very dirty, use a suitable proprietary cleaner with a brush to remove the ingrained dirt. Ideally, it's best to remove the carpets for cleaning, but this is an involved task in most modern cars, and it will probably be easier to leave them in place.

The facia, door trim, seats, and any other items which require attention can now be wiped over with a cloth soaked in warm water containing a little washing up liquid. If the trim is particularly dirty, use one of the proprietary cleaners available from motor accessory shops – a number of different types are available, suitable for vinyl, cloth or leather upholstery, etc, as required. An old nail brush or tooth brush will help to remove any ingrained dirt. On completion, wipe the surfaces dry using a lint-free cloth, and leave the windows open to speed up drying.

The inside of the windscreen and windows should be cleaned using a proprietary glass cleaner, or methylated spirits. Be careful about using certain household products such as washing-up liquid which may leave a smeary film. Finish off by wiping with a clean, dry paper tissue.

Don't forget to clean the boot.

Check for any nicks or tears in the interior trim, seats, headlining, etc. Various repair kits are available from motor accessory shops to suit most types of trim, but if the damage is very serious, the relevant trim panel will probably have to be renewed.

CLEANING THE BODYWORK

Ideally, the car should be washed every week, either by hand (preferably using a hosepipe), or by using a local car-wash. If you're washing the car by hand, use plenty of water to loosen the dirt and dust, and if possible use a suitable car shampoo or wax additive in the water. Any stubborn dirt such as road tar or bird droppings can be removed using methylated spirit or preferably one of the special proprietary cleaning solutions (in which case make sure that the product is suitable for use on car paint) – in either case, wash the affected area down with plenty of water after removing the dirt. **Never** just wipe over a dirty car, as this will scratch the paint very effectively.

FORD FIESTA

BODYWORK AND INTERIOR CARE

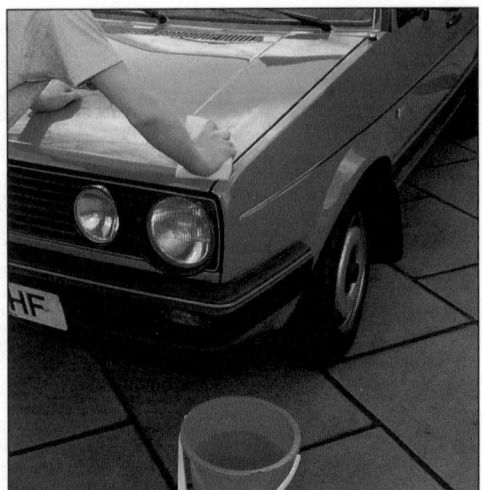

Two or three times a year, a good silicone or wax polish can be used on the paintwork. It's important to wash the car and remove all stubborn dirt, tar, etc, before using polish, as the polish will effectively seal over the top of any dirt, making it extremely difficult to remove in the future. Good polishes actually form a protective coating over the paint finish, which should help to make future accumulated dirt easier to remove. Always follow the manufacturer's recommendations closely when using polish. Try to avoid getting polish on any unpainted plastic or rubber body parts such as bumpers and spoilers, as it tends to leave a stain when dry. Chrome parts are best cleaned with a special chrome cleaner, as ordinary metal polish will wear away the finish.

Plastic or rubber body parts such as bumpers and spoilers should be cleaned using a suitable proprietary cleaner. A number of different types are available, including colour restorers, and special solvents to remove any stray polish which may have crept onto the plastic when polishing the paintwork. Make absolutely sure that any solvents or cleaners used are suitable, as certain products may attack plastic and/or rubber.

If the paint is beginning to lose its gloss or colour, and ordinary polishing doesn't seem to solve the problem, it's worth considering the use of a polish with a mild 'cutting' action to remove what is in effect a surface layer of 'dead' paint. In this case, follow the manufacturer's instructions, and don't polish too vigorously, or you might remove more paint than you intended!

DEALING WITH SCRATCHES

With superficial scratches which don't penetrate down to the metal, repair can be very simple.

Very light scratches can be polished out by carefully using a suitable 'cutting' polish. Follow the manufacturer's instructions, and take care not to remove too much of the surrounding paint.

If the scratch cannot be polished out, touch-up paint will be required. Touch-up paint is usually available in the form of a touch-up stick from the car manufacturer's dealers, or in various forms from motor accessory shops. You will probably have to quote the year and model of the car in order to make sure that you obtain the correct matching colour, and in some cases you may have to quote a paint reference number which will usually appear on the car's Vehicle Identification Number (VIN) plate under the bonnet (refer to *Servicing* on pages 82 & 83).

Lightly rub the area of the scratch with a very fine cutting paste to remove loose paint from the scratch and to clear the surrounding bodywork of polish.

Rinse the area with clean water.

Apply suitable touch-up paint to the scratch using a fine paint brush, and continue to apply fine layers of paint until the surface of the paint in the scratch is level with the surrounding paintwork.

Allow the new paint at least two weeks to harden, then blend it into the surrounding paintwork by rubbing the scratch area with a paintwork renovator or a very fine cutting paste.

If the paint finish requires a lacquer coat (in which case the lacquer will be supplied with the touch-up paint), it should now be applied to the newly painted area and allowed to dry in accordance with the manufacturer's instructions.

Finally, apply a suitable wax polish to the affected area for added protection.

FORD FIESTA

BULB, FUSE AND RELAY RENEWAL

BULBS

A defective exterior light can be not only dangerous, but also illegal. Carrying spare bulbs will enable you to renew blown ones as they occur. A failed interior light bulb may be just a nuisance, but a faulty exterior light could be a life-or-death matter.

Before assuming that a bulb has failed, check for corrosion on the bulbholder and wiring connections, particularly around the rear light assemblies.

- Note that as a safety precaution, the battery earth (black) lead should always be disconnected before renewing a bulb.
- Remember to reconnect the lead on completion.

Bulb ratings

Always make sure that when a bulb is renewed, a new bulb of the correct rating is used.

All the bulbs are of the 12-volt type, and their ratings are as follows:

Bulb	Rating [watts]
Headlight	60/55
Headlight (tungsten) – 'Mark 1' models	50/45
Headlight (tungsten) – 'Mark 2' models	45/40
Headlight (halogen)	60/55
Front sidelight ('Mark 1' models)	4
Front sidelight ('Mark 2' models)	5
Front and rear direction indicator lights	21
Front driving light	55
Front direction indicator repeater light	5
Rear brake/tail light	21/5
Reversing light	21
Rear foglight	21
Rear number plate light ('Mark 1' models)	5
Rear number plate light ('Mark 2' models)	10
Courtesy light	10
Boot light	10
Glovebox light ('Mark 1' models)	2
Glovebox light ('Mark 2' models)	4

FORD FIESTA

108 BULB, FUSE AND RELAY RENEWAL

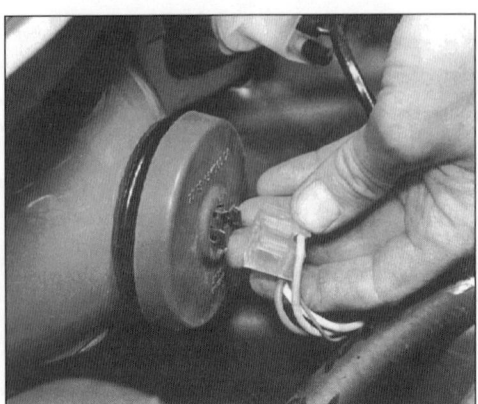

▲ Pulling the wiring plug from the back of a headlight ('Mark 2' model)

▲ Removing the rubber cover from the back of a headlight ...

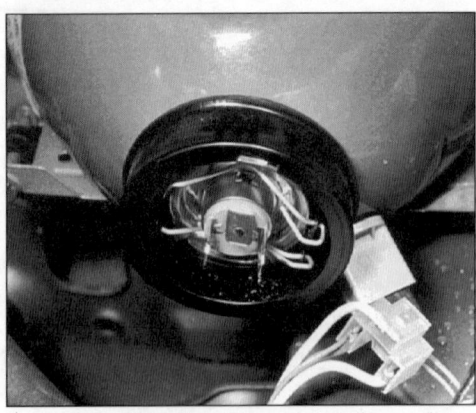

▲ ... to expose the bulb securing spring clip (arrowed) – 'Mark 2' model

Exterior light bulbs

HEADLIGHT

Open the bonnet. Working in the engine compartment, pull the wiring plug from the back of the headlight.

Pull the rubber cover from the back of the headlight, then rotate the bulb securing clip, or release the spring clip arms from bulb housing, according to type, and withdraw the bulb. Note that the bulb may be either a tungsten or halogen type – tungsten bulbs have a conventional spherical type glass bubble, whereas halogen types have a cylindrical glass

▲ Removing a headlight bulb (halogen type) – 'Mark 2' model

▲ Headlight assembly ('Mark 1' models)
1 Wiring plug
2 Rubber cover
3 Bulb securing clip
4 Bulb (tungsten type)
5 Sidelight bulbholder

FORD FIESTA

BULB, FUSE AND RELAY RENEWAL

section with a tapered point at the end. Fit the new bulb, taking care not to touch it with your fingers (hold the bulb using a clean cloth or paper tissue), using a reversal of the removal procedure. If the glass on the new bulb is inadvertently touched, wipe it with a cloth moistened with methylated spirit.

Make sure that the rubber cover is securely fitted to the back of the headlight.

FRONT SIDELIGHT

Open the bonnet, then remove the sidelight bulbholder from the side of the headlight by twisting it anti-clockwise.

Withdraw the push-fit bulb from the holder.

Fit the new bulb using a reversal of the removal procedure.

▲ Removing a front direction indicator light bulb ('Mark 1' models)

▲ Removing a front direction indicator light bulb ('Mark 2' models). Twist the bulbholder **(A)** and remove the bulb **(B)**

'Mark 1' XR2 models

The indicator lights are mounted in the front bumper.

Unscrew the two lens retaining screws from the front of the light assembly, and lift off the

▲ Removing a sidelight bulbholder ('Mark 2' model)

FRONT DIRECTION INDICATOR LIGHT

All models except 'Mark 1' XR2

Open the bonnet. On 'Mark 1' models, the direction indicators are accessible from the engine compartment below the headlights. On 'Mark 2' models, the direction indicators are mounted next to the headlights.

Twist the bulbholder anti-clockwise and pull it from the reflector. Remove the bulb by pushing it down into the bulbholder and turning it anti-clockwise.

Fit the new bulb and refit the bulbholder using a reversal of the removal procedure.

▲ Front direction indicator assembly ('Mark 1' XR2 models)
1 Lens retaining screw   3 Washer
2 Lens   4 Bulb

FORD FIESTA

BULB, FUSE AND RELAY RENEWAL

lens. Take care not lose any washers or rubber seals which may be fitted to the lens retaining screws.

Remove the bulb by pushing it down into the light and turning it anti-clockwise.

Fit the new bulb and refit the lens using a reversal of the removal procedure.

FRONT DRIVING LIGHT

Remove the retaining screw at the bottom of the lens, and withdraw the lens assembly from the light.

Disconnect the wires from the connectors inside the light housing, and remove the lens assembly from the car.

Release the spring clip arms from the back of the lens assembly, and withdraw the bulb (which is an 'H3' type, with integral bulbholder and wire).

Fit the new bulb, taking care not to touch it with your fingers (hold the bulb using a clean cloth or paper tissue), using a reversal of the removal procedure. If the glass on the new bulb is inadvertently touched, wipe it with a cloth moistened with methylated spirit.

Refit the lens assembly, making sure that the wires are securely connected.

FRONT DIRECTION INDICATOR REPEATER LIGHT

'Mark 1' models

Reach up behind the front wheelarch and locate the back of the repeater light bulbholder.

Pull the bulbholder from the rear of the light assembly, taking care not to strain the wiring.

Remove the bulb by pushing it down into the bulbholder and turning it anti-clockwise.

Fit the new bulb, then refit the bulbholder using a reversal of the removal procedure.

'Mark 2' models

Reach up behind the back of the front wheelarch and locate the back of the repeater light holder.

Squeeze the two clips on the back of the holder body, and push the assembly out through the wing panel.

Twist the bulbholder anti-clockwise to remove it from the lens, then pull the bulb from its socket.

Push a new bulb into the socket, and refit the assembly in the reverse order to removal.

REAR BRAKE/TAIL LIGHT AND REAR DIRECTION INDICATOR LIGHT ('MARK 1' MODELS)

Open the tailgate. Where applicable, lift up the floor mat, then remove the relevant cover panel from the floor of the boot to enable access to the rear light assembly. On later models, the cover panel may be secured with a clip. The clip must be lifted and turned, so that it can pass through the slot in the cover panel, before the cover panel can be lifted. Lift the cover panel over the clip, and pull it (towards the back of the car) from the retaining tabs.

The trim panel must be removed from the rear of the light assembly for access to the bulbs. Two alternative types of trim panel may be fitted depending on model.

On some models, the trim panel is removed by prising the retaining clip from its location in the body panel. To do this, carefully slide a

▲ Front driving light lens retaining screw (A)

▲ Removing a front driving light bulb

FORD FIESTA

BULB, FUSE AND RELAY RENEWAL

wide-bladed screwdriver (or a similar tool) behind the lower edge of the trim panel, and gently lever the centre of the panel until the retaining clip is released. Pull the trim panel down and away from the side of the boot to remove it.

On certain other models, the trim panel can be folded back for access to the bulbs. The panel is secured by a plastic retainer at the bottom corner, nearest the back of the car. To release the retainer, insert a coin into the slot and turn until the slot is vertical, then carefully fold the panel back to expose the light assembly.

To renew a bulb, squeeze the clips at the top and bottom of the relevant bulbholder, then pull the bulbholder from the back of the light assembly.

Remove the bulb by pushing it down into the bulbholder and turning it anti-clockwise.

Fit the new bulb, and refit the trim panel and the floor cover panel using a reverse of the removal procedure.

REAR BRAKE/TAIL LIGHT, REAR DIRECTION INDICATOR LIGHT AND REVERSING LIGHT ('MARK 2' MODELS)

Open the tailgate, then remove the relevant cover panel from the floor of the boot to enable access to the rear light assembly. The cover panel is secured by a clip, which must be lifted and turned, so that it can pass through the slot in the cover panel, before the cover panel can be lifted. Lift the cover panel over the clip, and pull it (towards the back of the car) from the retaining tabs.

The trim panel at the rear of the light assembly must be folded back for access to the bulbs. The panel is secured by a plastic retainer at the bottom corner, nearest the back of the car. To release the retainer, insert a coin into the slot and turn until the slot is vertical, then carefully fold the panel back to expose the light assembly.

Push the retaining clip on the back of the bulbholder upwards, and withdraw the complete bulbholder through one of the

▲ Rear brake/tail light and rear direction indicator light bulb renewal ('Mark 1' models)
1 Trim panel plastic retainer
2 Squeeze the bulbholder retaining clips (arrowed)

▲ Removing a rear direction indicator light bulb ('Mark 1' model). Trim panel retainer hole arrowed

▲ Rear brake/tail light, rear direction indicator light and reversing light bulb renewal ('Mark 2' models)
A Trim panel plastic retainer
B Bulbholder retaining clip

FORD FIESTA

BULB, FUSE AND RELAY RENEWAL

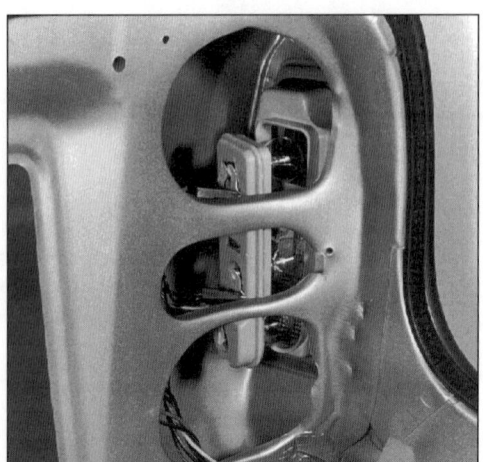

▲ *Rear light bulbholder released from body panel ('Mark 2' model)*

cut-outs in the body panel.

To remove a bulb, push it down into the bulbholder and turn anti-clockwise.

Fit the new bulb, and refit the bulbholder, ensuring that the retaining clip is correctly engaged.

Refit the trim panel and the floor cover panel using a reverse of the removal procedure.

REVERSING LIGHT ('MARK 1' MODELS)

Open the tailgate. Where applicable, lift up the floor mat, then remove the left-hand cover panel from the floor of the boot to enable access to the reversing light assembly. On later models, the cover panel may be secured with a clip. The clip must be lifted and turned, so that it can pass through the slot in the cover panel, before the cover panel can be lifted. Lift the cover panel over the clip, and pull it (towards the back of the car) from the retaining tabs.

On some models, it may be necessary to remove the spare wheel to improve access to the back of the reversing light (refer to 'Changing a wheel' in *'Breakdowns'* on page 38).

Pull the bulbholder from the back of the light assembly.

To remove the bulb, push it down into the bulbholder and turn anti-clockwise.

Fit the new bulb, then refit the bulbholder,

spare wheel (where applicable) and boot floor cover panel, using a reversal of the removal procedure.

REAR FOGLIGHT

Open the tailgate. Where applicable, lift up the floor mat, then remove the relevant cover panel from the floor of the boot to enable access to the foglight assembly. On later models, the cover panel may be secured with a clip. The clip must be lifted and turned, so that it can pass through the slot in the cover panel, before the cover panel can be lifted. Lift the cover panel over the clip, and pull it (towards the back of the car) from the retaining tabs. Where applicable, remove the storage box from the right-hand boot space after unscrewing the two plastic retaining screws, or remove the spare wheel from the left-hand boot space (refer to 'Changing a wheel' in *'Breakdowns'* on page 38).

▲ *Rear foglight securing nuts (arrowed) – 'Mark 2' model*

▲ *Removing the rear foglight ('Mark 2' model)*

FORD FIESTA

BULB, FUSE AND RELAY RENEWAL

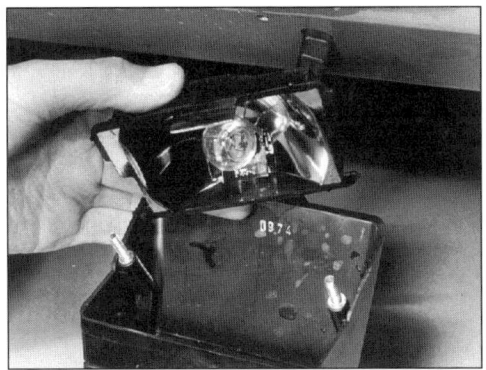

▲ Removing the rear foglight reflector ('Mark 2' model)

Working in the boot space, disconnect the foglight wiring plug.

Unscrew the two nuts or screws (as applicable) securing the foglight assembly to the body panel, then working outside the car, remove the assembly from the back of the car.

Working at the back of the light, press the retaining lugs at each side of the reflector, and remove the reflector.

To remove the bulb, push it down into the reflector and turn anti-clockwise.

Fit the new bulb, then reassemble and refit the foglight using a reversal of the removal and dismantling procedure.

REAR NUMBER PLATE LIGHT

'Mark 1' models

Reach up under the rear bumper, and locate the back of the number plate light.

Squeeze the clips at each end of the light assembly, and push the assembly out through the top of the bumper.

Unclip the light cover and the lens, then remove the bulb by pushing it into the light assembly and turning anti-clockwise.

Fit the new bulb, then reassemble and refit the light using a reversal of the removal and dismantling procedure.

'Mark 2' models

Using a screwdriver or similar tool, carefully prise the light assembly from the top of the bumper.

Squeeze the clips at each end of the light assembly, and remove the lens.

Remove the bulb by pushing it into the light assembly and turning anti-clockwise.

Fit the new bulb, then reassemble and refit the light using a reversal of the removal and dismantling procedure.

▲ Prising the rear number plate assembly from the bumper ('Mark 2' model)

▲ Rear number plate light assembly ('Mark 1' models)
  1 Retaining clips 2 Cover 3 Lens

▲ Removing the rear number plate light lens to expose the bulb ('Mark 2' model)

FORD FIESTA

BULB, FUSE AND RELAY RENEWAL

Interior light bulbs

Note: *Renewal of the various instrument panel and facia-mounted control illumination light bulbs requires detailed explanation, and in some cases extensive dismantling. Details can be found in the relevant Owners Workshop Manual.*

COURTESY LIGHT
Carefully prise the light assembly from its location in the roof, taking care not to damage the roof trim, then turn the bulb and pull it from the spring contacts in the light body. Fit the new bulb, making sure that it locates securely in the spring contacts, then refit the light assembly.

▲ *Push out the glovebox light assembly **(A)** and remove the bulb **(B)** – 'Mark 2' models*

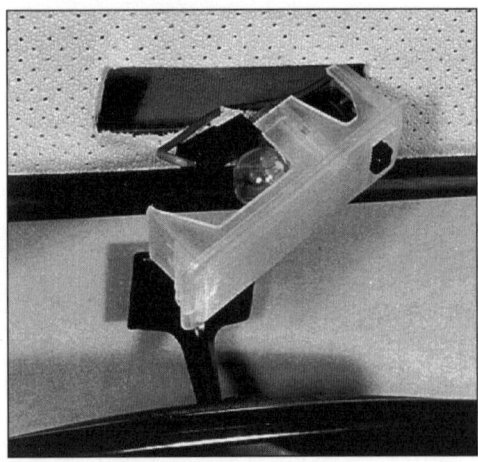

▲ *Removing the courtesy light ('Mark 2' model)*

BOOT LIGHT
Using a small screwdriver, carefully prise the light assembly from its location, then pull the bulb from the spring contact clip.

Fit the new bulb, making sure that it locates securely in the spring contact clip, then refit the light assembly.

GLOVEBOX LIGHT
Open the glovebox lid.

On 'Mark 1' models, simply pull the bulb from its socket in the side of the glove compartment.

On 'Mark 2' models, push the light assembly forwards out of the latch unit. To remove the bulb, push it gently into the bulbholder, and twist anti-clockwise.

Fit the new bulb using a reversal of the removal procedure.

FUSES

Most of the car's electrical circuits are protected by fuses which will blow when the relevant circuit becomes overloaded. This is to prevent possible damage to electrical components, or the possible risk of fire.

- Before renewing a fuse, always make sure that the ignition and the relevant circuit are switched off.

'Mark 1' models
The fuses are located behind a plastic cover panel under the facia at the top of the driver's oddments tray.

▲ *Fuse location (arrowed) – 'Mark 1' models*

FORD FIESTA

BULB, FUSE AND RELAY RENEWAL

The fuses are numbered (the numbers are marked on the fusebox) to identify the circuit which they protect.

To check a fuse, carefully prise the fusebox cover panel from the facia to expose the fuses.

Note that the fuse for the heated rear window circuit is mounted on top of the heated rear window relay, and can be renewed after removing the cover from the top of the relay (refer to the following Section *'Relays'* on page 116 for details of relay locations).

▲ *Removing the cover from the heated rear window relay to expose the fuse ('Mark 1' models)*

Conventional glass type fuses are used. To remove a fuse, push the retaining spring clips apart and withdraw the fuse. A blown fuse can be recognised by a break in the wire inside the glass envelope.

The fuse ratings are normally marked clearly on the fuses, either inside the glass envelope, or on the metal ends of the fuse. For reference, the fuse ratings are given in the following tables:

FUSE RATINGS FOR 'MARK 1' MODELS UP TO 1982

Fuse number	Rating (amps)
1	8
2	16
3	8
4	8
5	8
6	8
7	8

FUSE RATINGS FOR 'MARK 1' MODELS FROM 1982

Fuse number	Rating (amps)
1	8
2	8
3	8
4	16
5	8
6	8
7	8
8	8
11 or 12	16

If a fuse has blown, renew it with one of identical rating. Always use proper fuses of the correct type and rating, and **never** use wire or any other material to bridge the gap where a fuse should be.

When fitting a new fuse, clip the fuse into position so that the fuse wire is visible.

Note that the same fuse should never be renewed more than once without investigating the source of the trouble. Seek advice from someone suitably qualified if necessary.

The radio has its own in-line circuit fuse (usually located in the wiring loom behind the radio), or may be protected by a fuse in the rear of the radio casing.

'Mark 2' models

The fuses are located behind a plastic cover panel in the facia at the back of the driver's oddments tray.

The fuses are numbered (the numbers are stamped into the fusebox above each fuse) to identify the circuit which they protect, and the circuits are represented by symbols on the plastic fusebox cover.

To check a fuse, carefully prise the fusebox cover panel from the facia to expose the fuses.

The fuses are a push fit in their sockets, and a plastic tool is attached to the inside of the fusebox cover panel for use when changing fuses. To remove a fuse, push the tool over the end of the fuse, then use the tool to pull the fuse from its socket. A blown fuse can be recognised by a break in the wire between the two terminals.

BULB, FUSE AND RELAY RENEWAL

▲ Fusebox ('Mark 2' models)
1 Fusebox cover
2 Fuse wire
3 Fuse removal tool

The fuses are colour-coded to show their rating as follows:

FUSE RATINGS FOR 'MARK 2' MODELS

Fuse number	Rating (amps)
Red	10
Blue	15
Yellow	20
Clear	25
Green	30

Spare fuses are attached to the fusebox lid.

If a fuse has blown, renew it with one of identical rating. Always use proper fuses of the correct type and rating, and **never** use wire or any other material to bridge the gap where a fuse should be.

To fit a fuse, simply push it into the relevant socket.

Note that the same fuse should never be renewed more than once without investigating the source of the trouble. Seek advice from someone suitably qualified if necessary.

The radio has its own in-line circuit fuse (usually located in the wiring loom behind the radio), or may be protected by a fuse in the rear of the radio casing.

RELAYS

Various types of relays are used in some circuits, and a faulty relay may prevent one or more components from working.
● Before renewing a relay, always make sure that the ignition and the relevant circuit are switched off.

'Mark 1' models

The relays for the heated rear window, intermittent wipe, and direction indicator/hazard flasher circuits are of the plug-in type, and are mounted on a bracket located behind the facia on the driver's side.

For access to the relays, it's necessary to remove the lower trim panel from the driver's side facia. The panel is secured by two screws.

With the trim panel removed, reach up under the facia, disconnect the wiring plug from the relevant relay, and unclip the relay from the bracket.

▲ Relays mounted on bracket behind facia ('Mark 1' models)
A Heated rear window
B Intermittent wipe
C Direction indicators/hazard flashers

Refitting is a reversal of removal.

Various additional relays may be used depending on the equipment fitted, and these may be located in various positions depending on model. If a problem is suspected which may be due to a relay not located on the bracket behind the facia, it's best to seek advice from a Ford dealer who will know where on the car to

BULB, FUSE AND RELAY RENEWAL 117

find the relevant components.

It's difficult to tell visually whether a relay is faulty, and the best test is to substitute a known good relay to see if the relevant circuit then operates – if it does, then it's probably the relay which was causing the problem.

'Mark 2' models

The relays are of the plug-in type, and are mounted in the rear of the fusebox.

For access to the relays it's necessary to lower the fusebox from the facia as follows.

Carefully release the plastic retaining clips (which are a push-fit), and pull the lower trim panel ('carpet') from under the driver's side facia.

Prise the fusebox cover panel from the facia to expose the fuses and the three fusebox securing screws.

Remove the three securing screws and carefully lower the fusebox from behind the facia.

▲ Fusebox removed for access to relays ('Mark 2' models)

▲ Fusebox securing screws ('Mark 2' models)

Release the securing clips from the corners of the relay cover, and remove the cover.

To remove a relay, simply pull it from its socket.

The relay locations are shown in the accompanying illustration.

Various additional relays may be used depending on the equipment fitted, and these may be located in various positions depending on model. If a problem is suspected which may be due to a relay not located in the fusebox, it's best to seek advice from a Ford dealer who will know where on the car to find the relevant components.

It's difficult to tell visually whether a relay is faulty, and the best test is to substitute a known good relay to see if the relevant circuit then operates – if it does, then it's probably the relay which was causing the problem.

▲ Relay locations ('Mark 2' models)
- A Relay cover
- I Ignition
- II Heated rear window
- III Remote tailgate release (models up to 1986), CTX transmission, or headlight dipped beam (manual gearbox models from 1986)
- IV Intermittent wipe
- V Headlight dim-dip (models from 1986, alternative location)
- VI Direction indicators/hazard flashers
- VII Not used
- VIII Headlight main beam
- IX Headlight dipped beam (CTX transmission models)
- X Headlight dim-dip (models from 1986, alternative location)

FORD FIESTA

118 PREPARING FOR THE MOT TEST

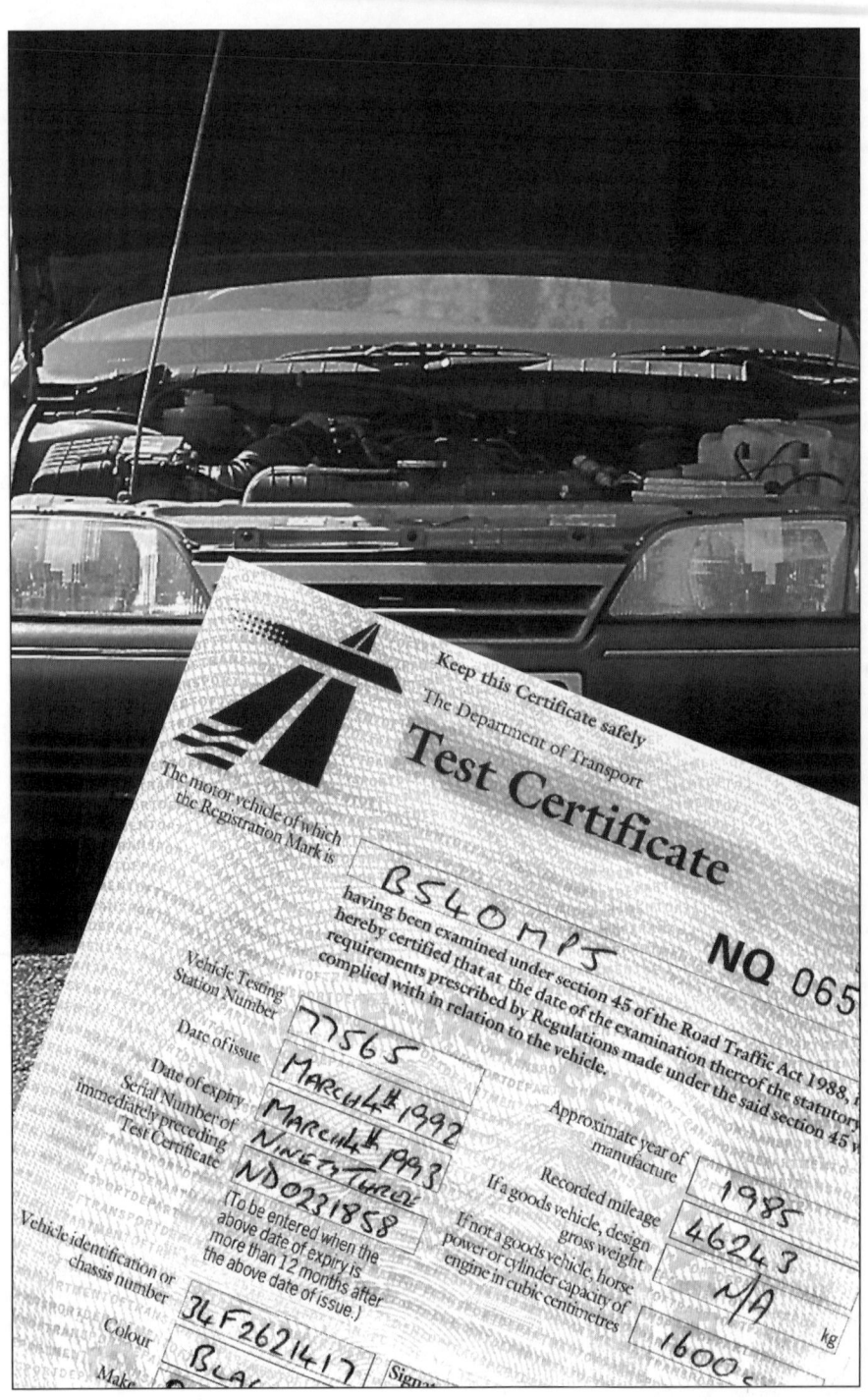

FORD FIESTA

PREPARING FOR THE MOT TEST

The following information is intended as a guide to enable you to spot some of the more obvious faults which may cause your car to fail the MOT test.

Obviously it isn't possible to check the car to the same standard as a professional MOT tester, who will be highly experienced and will have all the necessary tools and equipment.

Although we can't cover all the points of the test, the following should provide you with a good indication as to the general condition of the car, and will enable you to identify any obvious problem areas before submitting your car for the test.

Further explanation of most of the checks, and details of how to cure any problems discovered can be found in the relevant *Owners Workshop Manual* for your particular car (OWM 334 for Fiesta 'Mark 1' models, or OWM 1030 for Fiesta 'Mark 2' models).

Lights

- All lights must work, and the lenses and reflectors must not be damaged.
- Pairs of similar lights must be of the same brightness (eg both rear lights).
- Both headlights must show the same colour, and must be correctly aimed so that they light up the road adequately, but don't dazzle other drivers.
- The brake lights must work when the brake pedal is pressed.
- The direction indicator lights must flash between one and two times per second.

Steering and suspension

- All the components and their mounting points must be secure, and there must be no damage or excessive corrosion.
- There must be no excessive free play in the joints and bushes, and all rubber gaiters must be secure and undamaged.
- The steering mechanism must operate smoothly without excessive free play or roughness.
- There should be no excessive free play or roughness in the wheel bearings, but there should be sufficient free play to prevent tightness or binding.
- There should be no major fluid leaks from the shock absorbers and, when each corner of the car is depressed, it should rise and then settle in its normal position.
- The driveshafts and propeller shaft (on rear wheel drive cars) must not be damaged or distorted.

Brakes

- It should be possible to operate the handbrake without excessive force or excessive movement of the lever, and it must not be possible for the lever to release unintentionally. The handbrake must be able to lock the relevant wheels on which it operates.
- The brake pedal must not be damaged, and when the pedal is pressed, resistance should be felt near the top of its travel – hard resistance should be felt, and the pedal should not move down towards the floor. The resistance should be firm and not spongy.
- There should be no signs of any fluid leaks anywhere in the braking system.
- The wheels should turn freely when the brakes are not applied.
- When the brakes are applied, the brakes on all four wheels must work, and the car must stop evenly in a straight line without pulling to one side.
- All the braking system components must be secure, and there must be no signs of excessive wear or corrosion.

PREPARING FOR THE MOT TEST

Tyres and wheels

- The tyres must be in good condition, and there must be no signs of excessive wear or damage (refer to *Regular checks* on page 68).
- Tyres at the same end of the car must be of the same size and type. Always fit radial tyres; crossply tyres are not suitable for modern cars.

Seatbelts

- The mountings must be secure, and must not be loose or excessively corroded.
- The seatbelt webbing should not be frayed or damaged (this must be checked along the full length of the belts).
- When a seatbelt is fastened, the locking mechanism should hold securely and should release when intended.
- In the case of inertia reel seatbelts, the retractors should work properly when the belts are released.

General

- The windscreen wipers and washers must work properly, and the wiper blades must clear the windscreen without smearing.
- The horn must work properly.
- The exhaust mountings must be secure, and the system itself must be free from leaks and serious corrosion.
- There must be no serious corrosion or damage to the vehicle structure. Corrosion of body panels will not necessarily cause a car to fail the test, but there should be no serious corrosion of any of the 'load bearing' structural components.
- The exhaust gas emissions level must be within certain limits depending on the age of the car (special equipment is required to measure exhaust gas emissions), and the MOT test also includes a visual check for excessive exhaust smoke when the engine is running. Generally, the car should meet the emission regulations if regular servicing has been carried out.

FORD FIESTA

FAULT FINDING

The following table isn't intended as an exhaustive guide to fault finding, but it summarises some of the more common faults which may crop up during a car's life.

Hopefully, the table should help you to find the cause of a problem, even if you can't cure it yourself.

When confronted with a fault, try to think calmly and logically about the symptom(s), and you should be able to work out what the fault can't be! Check one item at a time, otherwise if you do clear the fault, you may not know what was causing it.

The commonest cause of difficulty is starting, especially in the Winter. Make sure that your battery is kept fully charged, and that the ignition components (ignition coil, distributor cap, spark plugs, HT leads, etc) are in good condition, clean and dry (refer to *'Servicing'* on page 85).

Further details of fault diagnosis, along with comprehensive renewal procedures for most components can be found in our *Owners Workshop Manual* for your particular model (OWM 334 for Fiesta 'Mark 1' models, or OWM 1030 for Fiesta 'Mark 2' models).

SYMPTOM	POSSIBLE CAUSES
ENGINE	
Starter motor doesn't turn, and headlights don't come on	• Flat battery • Loose, dirty or corroded battery connections • Other electrical or wiring fault
Starter motor doesn't turn, and headlights dim	• Battery charge low • Loose, dirty or corroded battery connections • Faulty starter motor • Other electrical or wiring fault • Seized engine
Starter motor doesn't turn, and headlights are bright	• Loose or dirty starter motor connections • Faulty starter motor or ignition switch • Gear selector lever not in position **N** or **P** (automatic transmission only) • Other electrical or wiring fault
Starter motor spins but doesn't turn engine	• Faulty starter motor • Engine fault
Engine turns slowly but won't start	• Battery charge low • Electrical or wiring fault • Wrong grade of engine oil (refer to *'Service specifications'* on page 61)

FAULT FINDING

SYMPTOM	POSSIBLE CAUSES
ENGINE (continued)	
Engine turns but won't fire, or engine starts but won't keep running	• Fuel tank empty • Damp or dirty ignition system components • Dirty or loose ignition system connections • Incorrectly adjusted or faulty spark plugs (refer to *'Servicing'* on page 86) • Fuel system fault • Other electrical or wiring fault
Engine idles, but stalls when accelerator pedal is depressed	• Blocked or dirty air cleaner filter element (refer to *'Servicing'* on page 97) • Fuel system fault • Ignition system fault • Electrical fault
Poor acceleration, misfiring or lack of power	• Incorrectly adjusted or faulty spark plugs (refer to *'Servicing'* on page 86) • Incorrect engine valve clearances (OHV engines only) • Fuel system fault • Ignition system fault • Electrical fault
Engine continues to run when ignition is switched off	• Engine overheating (possibly due to low coolant level – refer to *'Regular checks'* on page 67) • Incorrectly adjusted spark plugs or wrong type of spark plugs fitted (refer to *'Servicing'* on page 86) • Wrong grade of petrol • Fuel system fault • Excessive build-up of carbon inside engine ('decoke' required) • Faulty ignition switch • Other electrical or wiring fault
Engine doesn't reach normal operating temperature	• Faulty cooling system thermostat • Faulty temperature gauge or sensor

FAULT FINDING

SYMPTOM	POSSIBLE CAUSES
ENGINE (continued)	
Engine overheats or temperature gauge reads too high	• Airflow to radiator obstructed • Blocked radiator, hose or engine coolant passage • Coolant or engine oil level low (refer to 'Regular checks' on page 66) • Coolant hose(s) leaking, worn or damaged • Faulty cooling system thermostat • Faulty water pump • Faulty water pump drive • Faulty temperature gauge or sensor • Faulty electric cooling fan or switch
Ignition warning light comes on when engine is running	• Alternator drivebelt loose or broken (refer to 'Servicing' on page 95) • Faulty alternator • Other electrical or wiring fault
Oil pressure warning light comes on when engine is running	• Oil level below 'MIN' mark on dipstick – refer to 'Regular checks' on page 66 (if oil level is correct, and light is still on, seek advice, but **don't** start the engine) • Oil leak • Faulty oil pressure switch • Badly worn engine components
GEARBOX, TRANSMISSION AND CLUTCH	
Difficulty in engaging gear (manual gearbox)	• Worn or faulty gearbox components • Worn or faulty clutch mechanism • Engine idle speed too high
Clutch slips (car does not accelerate when engine revs increase – manual gearbox)	• Faulty clutch mechanism • Contaminated or worn clutch
Car pull away sluggish (automatic transmission)	• Low transmission fluid level (refer to 'Servicing' on page 92) • Fault in transmission
Engine speed too high, too low, or varies randomly when driving (automatic transmission)	• Fault in transmission

FAULT FINDING

SYMPTOM	POSSIBLE CAUSES
BRAKES	
Brakes feel 'spongy'	• Air in brake hydraulic system • Fluid leak in brake system
Excessive brake pedal travel	• Faulty rear brake mechanism • Air in brake hydraulic system • Fluid leak in brake system
Brakes require excessive pedal pressure	• Damp, dirty or contaminated brake components • Fluid leak in brake system • Faulty or seized brake components • Faulty brake servo (where applicable)
Car pulls to one side	• Incorrect tyre pressure(s) or uneven tyre wear (refer to *'Regular checks'* on page 68) • Faulty or seized brake components • Worn or contaminated brake friction material • Incorrect wheel alignment • Worn or faulty steering components • Worn or faulty suspension components
Brakes squeal and/or judder	• Badly worn or corroded brake components • Incorrectly assembled brake components • Contaminated brake friction material • Worn or faulty suspension components • Worn or faulty steering components
Brake fluid level warning light comes on (where applicable)	• Low brake fluid level (refer to *'Regular checks'* on page 68) • Faulty fluid level sensor or wiring

FORD FIESTA

FAULT FINDING

SYMPTOM	POSSIBLE CAUSES
SUSPENSION AND STEERING	
Car becomes heavy to steer	• Low tyre pressure(s) (refer to *'Regular checks'* on page 68) • Incorrect wheel alignment • Worn or faulty steering components • Worn or faulty suspension components
Car pulls to one side	• Refer to *'Brakes'* section of table
Car wanders	• Incorrect tyre pressure(s) or uneven tyre wear (refer to *'Regular checks'* on page 68) • Car unevenly loaded • Incorrect wheel alignment • Worn or faulty suspension components • Worn or faulty steering components
Car vibrates when driving	• Loose wheel bolt(s) • Wheel(s) out of balance • Worn or damaged driveshaft • Worn or faulty suspension components • Worn or faulty steering components • Worn or faulty brake components
Hard or choppy ride	• Incorrect tyre pressure(s) (refer to *'Regular checks'* on page 68) • Worn or faulty suspension components
Car leans excessively when cornering	• Roof rack overloaded • Car unevenly loaded • Worn or faulty suspension components
Uneven tyre wear	• Incorrect tyre pressure(s) (refer to *'Regular checks'* on page 68) • Incorrect wheel alignment • Wheel(s) out of balance • Worn or faulty suspension components • Worn or faulty steering components • Brakes 'grabbing'

FAULT FINDING

SYMPTOM	POSSIBLE CAUSES
ELECTRICS	
Electrical systems don't work	• Loose, dirty or corroded battery connections
	• Faulty earth connection
	• Battery charge low
	• Blown fuse (refer to *'Bulb, fuse and relay renewal'* on page 114)
	• Faulty relay
	• Fuse link, connecting main wiring loom to battery, blown (seek advice)
Direction indicators don't work	• Blown fuse (refer to *'Bulb, fuse and relay renewal'* on page 114)
	• Faulty earth connection
	• Faulty direction indicator relay
	• Faulty direction indicator switch
Bulbs burn out repeatedly	• Faulty connections at light socket
	• Faulty alternator regulator
All lights dim when engine speed drops to idle	• Loose alternator drivebelt (refer to *'Servicing'* on page 95)
	• Battery charge low
	• Faulty alternator

FORD FIESTA

CAR JARGON

This Section should help you to understand some of the odd words of phrases used at garages and by 'car enthusiasts' which you may not be familiar with.

A

ABS – Abbreviation for Anti-lock Braking System. Uses sensors at each wheel to sense when the wheels are about to lock, and releases the brakes to prevent locking. This process occurs many times per second, and allows the driver to maintain steering control when braking hard.
Accelerator pump – A device attached to many *carburettors* which provides a spurt of extra fuel to the carburettor fuel/air mixture when the accelerator pedal is suddenly pressed down.
Additives – Compounds which are added to petrol and oil to improve their quality and performance.
Advance and retard – A system for altering the *ignition timing*.
AF – An abbreviation of 'Across Flats', the way in which many nuts, bolts and spanners are identified. AF is usually preceded by an Imperial unit of measurement – eg ½in AF. Unless otherwise stated, all metric measurements are assumed to be AF, so the abbreviation is not normally used for metric nuts, bolts and spanners.
Air cooling – Alternative method of engine cooling in which no water is used. An engine-driven fan forces air at high speed over the surfaces of the engine.
ALB – Abbreviation for Anti-Lock Braking System. See *'ABS'*.
Alternator – A device for converting rotating mechanical energy into electrical energy. In modern cars, it has superseded the dynamo for charging the battery because of its much greater efficiency.
Ammeter – A device for measuring electrical current – the current supplied to the battery by the alternator, or drawn from the battery by the car's electrical systems.
Antifreeze – A chemical mixed with the water in the cooling system to lower the temperature at which the coolant freezes, and in modern cars to prevent corrosion of the metal in the cooling system.
Anti-roll bar – A metal bar mounted transversely across the car, connecting the two sides of the suspension, which counteracts the natural tendency for the car to lean when cornering.
Aquaplaning – A word used to describe the action of a tyre skating across the surface of water.
Automatic transmission – A type of *gearbox* which selects the correct gear ratio automatically according to engine speed and load.
Axle – Spindle on which a wheel revolves.

B

Balljoint – A ball-and-socket type joint used in steering and *suspension* systems, which allows relative movement in more than one plane.
Battery condition indicator – A device for measuring electrical voltage (a voltmeter) connected via the ignition switch to the car battery. Unlike an *ammeter*, it gives an indication that a battery is close to failing. **Also**, most 'maintenance-free' batteries have a battery condition indicator fitted to their casing, which consists of a small disc which changes its colour when the battery is close to failure and requires renewal.
Bearing – Metal or other hard wearing surface against which another part moves, and which is designed to reduce friction and wear (bearings are usually lubricated).

FORD FIESTA

CAR JARGON

Bendix drive – A device on some types of starter motor which allows the motor to drive the engine for starting, then disengages when the engine starts to run.
BHP – see *Horsepower*.
Big end – The end of a *connecting rod* which is attached to the *crankshaft*. It incorporates a *bearing* and transmits the linear movement of the connecting rod to the crankshaft.
Bleed nipple (or valve) – A hollow screw which allows air or fluid to be bled out of a system when it is loosened.
Brake caliper – The part of a *disc brake* system which houses the *brake pads* and the hydraulically-operated pistons.
Brake disc – A rotating disc, coupled to a roadwheel, which is clamped between hydraulically operated friction pads in a *disc brake* system.
Brake fade – A temporary loss of braking efficiency due to overheating of the brake friction material.
Brake pad – The part of a *disc brake* system which consists of the friction material and a metal backing plate.
Brake shoe – The part of a *drum brake* system which consists of the friction material and a curved metal former.
Breather – A device which allows fresh air into a system or allows contaminated air out.
Bucket tappet – A bucket shaped component used in some engines to transfer the rotary movement of the *camshaft* to the up-and-down movement required for *valve* operation.
Bump stop – A hard piece of rubber used in many *suspension* systems to prevent the moving parts from contacting the body during violent suspension movements.

C

Camber angle – The angle at which the front wheels are set from the vertical, when viewed from the front of the car. Positive camber is the amount in degrees which the wheels are tilted out at the top.
Cam follower – A piece of metal used to transfer the rotary movement of the *camshaft* to the up-and-down movement required for *valve* operation.
Camshaft – A rotating shaft driven from the *crankshaft* with lobes or cams used to operate the engine *valves* via the *valve gear*.
Carbon leads – Ignition HT leads incorporating carbon (black fibres) which eliminates the need for separate radio and TV *suppressors*.
Carburettor – A device which is used to mix air and fuel in the proportions required for burning by the engine under all conditions of engine running.
Castor angle – The angle between the front wheels pivot points and a vertical line when viewed from the side of the car. Positive castor is when the axis is inclined rearwards.
Catalytic converter – A device incorporated in the exhaust system which speeds up the natural decomposition of the exhaust gases, and reduces the amount of harmful gases released into the atmosphere. Cars fitted with catalytic converters must be operated on *unleaded petrol*, as *leaded petrol* will destroy the catalyst.
Centrifugal advance – System of ignition *advance and retard* incorporated in many *distributors* in which weights rotating on a shaft alter the ignition timing according to engine speed.
Choke – This has two common meanings. It is used to describe the device which shuts off some of the air in a *carburettor* during cold starting (in order to provide extra fuel), and it may be manually or automatically operated. It's also used as a general term to describe a carburettor throttle bore.
Clutch – A friction device which allows two rotating components to be coupled together smoothly, without the need for either rotating component to stop moving.
Coil spring – A spiral coil of spring steel used in many *suspension* systems.
Combustion chamber – Shaped area in the *cylinder head* into which the fuel/air mixture is compressed by the *piston* and where the spark from the *spark plug* ignites the mixture.
Compression ratio (CR) – A term used to describe the amount by which the fuel/air mixture is compressed as a *piston* moves from the bottom to the top of its travel, and expressed as a number. For example an 8.5:1 compression ratio means that the volume of fuel/air mixture above the piston when the piston is at the bottom of its stroke is 8.5 times

FORD FIESTA

CAR JARGON

that when the piston is at the top of its stroke.
Compression tester – A special type of pressure gauge which can be screwed into a *spark plug* hole, which measures the pressure in the cylinder when the engine is turning but not firing. This gives an indication of engine wear or possible leaks.
Condenser (capacitor) – A device in a *contact breaker point distributor* which stores electrical energy and prevents excessive sparking at the contact breaker points.
Connecting rod ('con-rod') – Metal rod in the engine connecting a *piston* to the *crankshaft*.
Constant velocity (CV) joint – A joint used in *driveshafts*, where the instantaneous speed of the input shaft is exactly the same as the instantaneous speed of the output shaft at any angle of rotation. This does not occur in ordinary *universal joints*.
Contact breaker points – A device in the *distributor* which consists of two electrical points (or contacts) and a cam which opens and closes them to operate the *HT* electrical circuit which provides the spark at the *spark plugs*.
Crankcase – The area of the *cylinder block* below the *pistons* which houses the *crankshaft*.
Crankshaft – A cranked shaft which is driven by the *pistons* and provides the engine output to the *transmission*.
Crossflow cylinder head – A *cylinder head* in which the inlet and exhaust *valves* and *manifolds* are on opposite sides.
Crossply tyre – A tyre whose construction is such that the weave of the fabric material layers is running diagonally in alternately opposite directions to a line around the circumference of the tyre.
Cubic capacity – The total volume within the *cylinders* of an engine which is swept by the movement of the *pistons*.
CVH – A term applied by the Ford Motor Company to their overhead camshaft engines which incorporate a hemispherical *combustion chamber*. CVH means Compound Valve angle, Hemispherical combustion chamber.
Cylinder – Close fitting metal tube in which a *piston* slides. In the case of an engine, the cylinders may be bored directly into the *cylinder block*, or on some engines, cylinder liners are used which rest in the cylinder block and can be replaced when worn with matching pistons to avoid the requirement for *reboring* the cylinder block.
Cylinder block – The main engine casting which contains the *cylinders*, *crankshaft* and *pistons*.
Cylinder head – The casting at the top of the engine which contains the *valves* and associated operating components.

D

Damper – See *shock absorber*.
Dashpot – An oil-filled *cylinder* and *piston* used as a damping device in SU and Zenith/Stromberg CD type carburettors.
Dead axle (beam axle) – The simplest form of axle, consisting of a horizontal member attached to the car underbody by springs. This arrangement is used for the rear axle on some front-wheel-drive cars.
Decarbonising ('decoking') – Removal of all carbon deposits from the *combustion chambers* and the tops of the *pistons* and *cylinders* in an engine.
De Dion axle – A rear axle consisting of a cranked tube attached to the wheel hubs, with a separately mounted *differential* gear and *driveshafts*. *Suspension* is normally through *coil springs* between the wheel hubs and car underbody.
Derv – Abbreviation for Diesel-Engined Road Vehicle. A term often used to refer to Diesel fuel.
Diaphragm – A flexible membrane used in some components such as fuel pumps. The diaphragm spring used on *clutches* is similar, but is made from spring steel.
Diesel engine – An engine which relies on the heat generated when compressing air to ignite the fuel, and which therefore doesn't need an *ignition system*. Diesel engines have much higher *compression ratios* than petrol engines, normally around 20:1.
Differential – A system of gears (generally known as a crownwheel and pinion) which allows the *torque* provided by the engine to be applied to both driving wheels. The differential divides the torque proportionally between the driving wheels to allow one wheel to turn faster than the other, for example during cornering.

CAR JARGON

DIN – This stands for Deutsche Industrie Norm (German Industry Standard), which provides international standards for measuring engine power, torque, etc.
Disc brake – A brake assembly where a rotating disc is clamped between hydraulically operated friction pads.
Distributor – A device used to distribute the HT current to the individual *spark plugs*. The distributor may also contain the *advance and retard* mechanism. On some older cars, the distributor also contains the *contact breaker points* assembly.
Distributor cap – Plastic cap which fits on top of the *distributor* and contains electrodes, in which the *rotor arm* rotates to distribute the HT *spark* voltage to the correct *spark plug*.
DOHC – Abbreviation for Double Overhead Camshaft (see *'Twin-cam'*).
Doughnut – A term used to describe the flexible rubber coupling used in some *driveshafts*.
Driveshaft – Term usually used to describe the shaft (normally incorporating *universal* or *constant velocity joints*), which transmits drive from a *differential* to one wheel. More commonly found in front-wheel-drive cars.
Drive train – A collective term used to describe the *clutch/gearbox/transmission* and the other components used to transmit drive to the wheels.
Drum brake – A brake assembly with friction linings on 'shoes' running inside a cylindrical drum attached to the wheel.
Dual circuit brakes – A *hydraulic* braking system consisting of two separate fluid circuits, so that if one circuit becomes inoperative, braking power is still available from the other circuit.
Dwell angle – A measurement which corresponds to the number of degrees of *distributor* shaft rotation during which the *contact breaker points* are closed during the ignition cycle of one *cylinder*. The angle is altered by adjusting the contact breaker points gap.

E

Earth strap – A flexible electrical connection between the battery and a car earth point, or between the engine/*gearbox* and the car body to provide a return current path flow to the battery.
EFI – Abbreviation for Electronic *Fuel Injection*.
Electrode – An electrical terminal, eg in a *spark plug* or *distributor cap*.
Electrolyte – A current-conducting solution inside the battery (consisting of water and sulphuric acid in the case of a car battery).
Electronic ignition – An *ignition system* incorporating electronic components in place of *contact breaker points*, which can produce a much higher spark voltage than a contact breaker system, and is less affected by worn components.
Emission control – The reduction or prevention of the release into the atmosphere of poisonous fumes and gases from the engine and fuel system of a car. Required to different degrees by the laws of some countries, and achieved by engine design and the use of special devices and systems.
Epicyclic gears (planetary gears) – A gear system used in many *automatic transmissions* where there is a central 'sun' gear around which smaller 'planet' gears rotate.
Exhaust gas analyser – An instrument used to measure the amount of pollutants (mainly carbon monoxide) in a car's exhaust gases.
Expansion tank – A container used in many cooling systems to collect the overflow from the car's cooling system as the coolant heats up and expands.

F

Filter – A device for removing foreign particles from air, fuel or oil.
Final drive – A collective term (often expressed as a gear ratio) for the crownwheel and pinion (see *Differential*).
Flat engine – Form of engine design in which the *cylinders* are opposed horizontally, usually with an equal number on each side of the central *crankshaft*.
Float chamber – The part of a *carburettor* which contains a float and *needle valve* for

CAR JARGON

controlling the fuel level in the reservoir.
Flywheel – A heavy rotating metal disc attached to the *crankshaft* and used to smooth out the pulsing from the *pistons*.
Four stroke (cycle) – A term used to describe the four operating strokes of a *piston* in a conventional car engine. These are (1) Induction – drawing the air/fuel mixture into the engine as the piston moves down; (2) Compression – of the fuel/air mixture as the piston rises; (3) Power stroke – where the piston is forced down after the fuel/air mixture has been ignited by the *spark plug*, and (4) Exhaust stroke – where the piston rises and pushes the burnt gases out of the *cylinder*. During these operations, the inlet and exhaust *valves* are opened and closed at the correct moment to allow the fuel/air mixture in, the exhaust gases out, or to provide a gas-tight compression chamber.
Fuel injection – A method of injecting fuel into an engine. Used in *Diesel engines* and also on some petrol engines in place of a *carburettor*.
Fuel injector – Device used on *fuel injection* engines to inject fuel directly or indirectly into the *combustion chamber*. Some fuel injection systems use a single fuel injector, while some systems use one fuel injector for each *cylinder* of the engine.

G

Gasket – Compressible material used between two surfaces to provide a leakproof joint.
Gearbox – A group of gears and shafts installed in a housing, positioned between the *clutch* and the *differential*, and used to keep the engine within its safe operating speed range as the speed of the car changes.

H

Half-shaft – A *driveshaft* used to transmit the drive from the *differential* to one of the rear wheels.
Hardy-Spicer joint (Hooke's or Cardan joint) – See *Universal joint*.
Helical gears – Gears in which the teeth are cut at an angle across the circumference of the gear to give a smoother mesh between gears and quieter running.
Horsepower – A measurement of power. Brake Horsepower (BHP) is a measure of the power required to stop a moving body.
HT – Abbreviation of High Tension (meaning high voltage) used to describe the *spark plug* voltage in an *ignition system*.
Hub carrier – A component usually found at each front corner of a car which carries the wheel and brake assembly, and to which the *suspension* and steering components are attached.
Hydraulic – A term used to describe the operation of a system by means of fluid pressure.
Hypoid gear – A gear with curved teeth which transmits drive through a right-angle, where the centreline of the drive gear is offset from the centreline of the driven gear. The meshing action of hypoid gears allows a larger and therefore stronger drive gear, and the meshing noise is reduced in comparison with conventional gears.

I

Independent suspension – A *suspension* system where movement of one wheel has no effect on the movement of the other, eg independent front suspension.
Ignition coil – An electrical coil which forms part of the *ignition system* and which generates the *HT* voltage.
Ignition system – The electrical system which provides the spark to ignite the air/fuel mixture in the engine. Normally the system consists of the battery, *ignition coil*, *distributor*, ignition switch, *spark plugs* and wiring.
Ignition timing – The time in the *cylinder* firing cycle at which the ignition spark (provided by the *spark plug*) occurs. The spark timing is normally a few degrees of *crankshaft* rotation before the *piston* reaches the top of its stroke, and is expressed as a number of degrees before top-dead-centre (BTDC).
Inertia reel – Automatic type of seat belt mechanism which allows the wearer to move freely in normal use, but which locks on sensing either sudden deceleration or a sudden movement of the wearer.

CAR JARGON

In-line engine – An engine in which the *cylinders* are positioned in one row as opposed to being in a *flat* or *vee* configuration.

J

Jet – A calibrated nozzle or orifice in a *carburettor* through which fuel is drawn for mixing with air.
Jump leads – Heavy electric cables fitted with clips to enable a car's battery to be connected to another battery for emergency starting.

K

Kerb weight – The weight of a car, unladen but ready to be driven, ie with enough fuel, oil, etc, to travel an arbitrary distance.
Kickdown – A device used on *automatic transmissions* which allows a lower gear to be selected for improved acceleration by fully depressing the accelerator.
Kingpin – A device which allows the front wheel of a car to swivel about a near vertical axis.
Knocking – See 'Pinking'.

L

Laminated windscreen – A windscreen which has a thin plastic layer sandwiched between two layers of toughened glass. It will not shatter or craze when hit.
Lead-free petrol – Contains no lead. It has no lead added during manufacture, and the natural lead content is refined out. This type of petrol is not currently available for general use in the UK, and should not be confused with *unleaded* petrol.
Leaded petrol – Normal 4-star petrol. Has a low amount of lead added during manufacture, in addition to the natural lead found in crude oil.
Leading shoe – A *drum brake* shoe of which the leading end (the one moved by the operating *pistons*) is reached first by a given point on the drum during normal forward rotation. A simple drum brake will have one leading and one trailing (the opposite) shoe.

Leaf spring – A spring commonly used on cars with a *live axle*, consisting of several long curved steel plates clamped together.
Limited slip differential – A type of *differential* which prevents one wheel from standing still while the other wheel spins excessively. Often used on high-performance cars.
Live axle – An axle through which power is transmitted to the rear wheels.
Loom – A complete car wiring system or section of a wiring system consisting of all the wires of correct length, etc, to wire up the various circuits.
LT – Abbreviation of Low Tension (meaning low voltage), used to describe battery voltage in the *ignition system*.

M

MacPherson strut – An independent front *suspension* system where the swivelling, springing and shock absorbing action of the wheels is dealt with by a single assembly.
Manifold – A device used for ducting the air/fuel mixture to the engine (inlet manifold), or the exhaust gases from the engine (exhaust manifold).
Master cylinder – A *cylinder* containing a *piston* and *hydraulic* fluid, directly coupled to a foot pedal (eg brake or *clutch* master cylinder). Used for transmitting pressure to the brake or clutch operating mechanism.
Metallic paint – Paint finish incorporating minute particles of metal to give added lustre to the colour.
Multigrade – Lubricating oil whose *viscosity* covers that of several single grade oils, making it suitable for use over a wider range of operating conditions.

N

Needle bearing – Type of *bearing* in which needle or cone-shaped rollers are used around the circumference to reduce friction.
Needle valve – A component of the *carburettor* which restricts the flow of fuel or fuel/air mixture according to the position of the valve in an orifice or *jet*.

CAR JARGON

Negative earth – Electrical system (almost universally adopted) in which the negative terminal of the car battery is connected to the car body. The polarity of all the electrical equipment is determined by this.

O

Octane rating – A scale rating for grading petrol.
ohc (overhead cam) – Describes an engine in which the *camshaft* is situated above the *cylinder head*, and operates the *valve gear* directly.
ohv (overhead valve) – Describes an engine which has its *valves* in the *cylinder head*, but with the *valve gear* operated by *pushrods* from a *camshaft* situated lower in the engine.
Oil cooler – Small *radiator* fitted in the oil circuit and positioned in a cooling airflow to cool the oil. Used mainly on high-performance engines.
Overdrive – A device coupled to a car's *gearbox* which raises the output gear ratio above the normal 1:1 of top gear.
Oversteer – A tendency for a car to turn more tightly into a corner than intended.

P

PCV (Positive Crankcase Ventilation) – A system which allows fumes and vapours which build up in the *crankcase* to be drawn into the engine for burning.
Pinion – A gear with a small number of teeth which meshes with one having a larger number of teeth.
Pinking – A metallic noise from the engine often caused by the *ignition timing* being too far *advanced*. The noise is the result of pressure waves which cause the *cylinder* walls to vibrate when the ignited fuel/air mixture is compressed.
Piston – Cylindrical component which slides in a closely-fitting metal tube or *cylinder* and transmits pressure. The pistons in an engine, for example, compress the fuel/air mixture, transmit the power to the *crankshaft*, and push the burnt gases out through the exhaust *valves*.

Piston ring – Hardened metal ring which is a spring fit in a groove running round the *piston* to ensure a gas-tight seal between the piston and *cylinder* wall.
Positive earth – The opposite of *negative earth*.
Power steering – A steering system which uses *hydraulic* fluid pressure (provided by an engine-driven pump) to reduce the effort required to steer the car.
Pre-ignition – See *'Pinking'*.
Propeller shaft – The shaft which transmits the drive from the *gearbox* to the rear axle in a front-engined rear-wheel-drive car.
Pushrod – A rod which is moved up and down by the rotary motion of the *camshaft* and operates the *rocker arms* in an *ohv* engine.

Q

Quarter light – A triangular window mounted in front or behind the main front or rear windows, usually in the front door, or behind the rear door.
Quartz-halogen bulb – A bulb with a quartz envelope (instead of glass), filled with a halogen gas. Gives a brighter, more even spread of light than an ordinary bulb.

R

Rack and pinion – Simplest form of steering mechanism which uses a *pinion* gear to move a toothed rack.
Radial ply tyre – A tyre in which the fabric material plies are arranged laterally, at right angles to the circumference.
Radiator – Cooling device through which the engine coolant is passed, situated in an air flow and consisting of a system of fine tubes and fins for rapid heat dissipation.
Radius arms (rods) – Locating arms sometimes used with a *live axle* to positively locate it in the fore-and-aft direction.
Rebore – The process of enlarging the *cylinder* bores to a very accurately specified measurement in order to fit new *pistons* to overcome wear in the engine. Not normally necessary unless the engine has covered a very high mileage.

CAR JARGON

Recirculating ball steering – A derivative of *worm and nut* steering, where the steering shaft motion is transmitted to the steering linkage by balls running in the groove of a worm gear.
Rev counter – See *Tachometer*.
Rocker arm – A lever which rocks on a central pivot, with one end moved up and down by the *camshaft*, and the other end operating an inlet or exhaust *valve*.
Rotary engine – See '*Wankel engine*'.
Rotor arm – A rotating arm in the *distributor* which distributes the *HT* spark voltage to the correct *spark plug*.
Running on – A tendency for an engine to keep on running after the ignition has been switched off. Often caused by a badly maintained engine or the use of an incorrect grade of fuel.

S

SAE – Society of Automotive Engineers (of America). Lays down international standards for the classification of engine performance and many other specifications, but is most commonly used to classify oils.
Safety rim – A special wheel rim shape which prevents a deflated tyre from rolling off the wheel.
Sealed beam – A sealed headlamp unit where the filament is an integral part and cannot be renewed separately.
Semi-trailing arm – A common form of independent rear *suspension*.
Servo – A device for increasing the normal effort applied to a control.
Shock absorber – A device for damping out the up-and-down movement of the *suspension* when the car hits a bump in the road.
Spark plug – A device with two *electrodes* insulated from each other by a ceramic material, which screws into an engine *combustion chamber*. When the *HT* voltage is applied to the plug terminal, a spark jumps across the electrodes and ignites the fuel/air mixture.
Squab – Another name for a seat cushion.
Steel-braced tyre – Tyre in which extra plies containing steel cords are incorporated with the fabric plies to give added strength.

Steering arm (knuckle) – A short arm on the front *hub carrier* to which the steering linkage connects.
Steering gear – A general term used to describe the steering components, usually refers to a steering rack-and-pinion assembly.
Steering rack – See *Rack and pinion*.
Stroboscopic light – A light switched on and off by the engine *ignition system* which is used for checking the *ignition timing* when the engine is running.
Stroke – The total distance travelled by a single *piston* in its *cylinder*.
Stub axle – A short axle which carries one wheel.
Subframe – A small frame which is mounted on the car's body, and carries the *suspension* and/or the *drivetrain* assemblies.
Sump – The main reservoir for the engine oil.
Supercharger – A device which uses an engine-driven turbine (usually driven by a belt or gears from the *crankshaft*) to drive a compressor which forces air into the engine, providing increased fuel/air mixture flow, and therefore increased engine efficiency. Sometimes used on high-performance engines.
Suppressor – A device which is used to reduce or eliminate electrical interference caused by the *ignition system* or other electrical components.
Suspension – A general term used to describe the components which suspend the car body on its wheels.
Swing axle – A *suspension* arm which is pivoted near the front-to-rear centreline of the car, and which allows the wheel to swing vertically about that pivot point.
Synchromesh – A device in a *gearbox* which synchronises the speed of one gear shaft with another to produce smooth, noiseless engagement of the gears.

T

Tachometer – Also known as a rev counter, indicates engine speed in revolutions per minute (rpm).
Tappet – A term often used to refer to the component which transmits the rotary *camshaft* movement to the up-and-down movement required for *valve* operation.

CAR JARGON

Thermostat – A device which is sensitive to changes in engine coolant temperature, and opens up an additional path for coolant to flow through the *radiator* (to increase the cooling) when the engine has warmed up.

Tie-rod – A rod which connects the *steering arms* to the *steering gear*.

Timing belt – Fabric or rubber belt engaging on sprocket wheels and driving the *camshaft* from the *crankshaft*.

Timing chain – Metal flexible link chain engaging on sprocket wheels and driving the *camshaft* from the *crankshaft*.

Timing marks – Marks normally found on the *crankshaft* pulley or the *flywheel* and used for setting the ignition firing point with respect to a particular *piston*.

Toe-in/toe-out – The amount by which the front wheels point inwards or outwards from the straight-ahead position when steering straight ahead.

Top Dead Centre (TDC) – The point at which a *piston* is at the top of its *stroke*.

Torque – The turning force generated by a rotating component.

Torque converter – A coupling where the driving *torque* is transmitted through oil. At low speeds there is very little transfer of torque from the input to the output. As the speed of the input shaft increases, the direction of fluid flow within a system of vanes changes, and torque from the input impeller is transferred to the output turbine. The higher the input speed, the closer the output speed approaches it, until they are virtually the same.

Torsion bar – A metal bar which twists about its own axis, and is used in some *suspension* systems.

Toughened windscreen – A windscreen which when hit, will shatter in a particular way to produce blunt-edged fragments or will craze over but remain intact. A zone toughened windscreen has a zone in front of the driver which crazes into larger parts to reduce the loss of visibility which occurs when toughened windscreens break, but is otherwise similar.

Track rod – See *Tie-rod*.

Trailing arm – A form of independent *suspension* where the wheel is attached to a swinging arm, and is mounted to the rear of the arm pivot.

Transaxle – A combined *gearbox*/axle assembly from which two *driveshafts* transmit the drive to the wheels.

Transmission – A general term used to describe some or all of the *drivetrain* components excluding the engine, most commonly used to describe automatic gearboxes.

Turbocharger – A device which uses a turbine driven by the engine exhaust gases to drive a compressor which forces air into the engine, providing increased fuel/air mixture flow, and therefore increased engine efficiency. Commonly used on high-performance engines.

Twin-cam – Abbreviation for twin overhead *camshafts* (see *'ohc'*). Used on engines with a *crossflow cylinder head*, usually with one camshaft operating the inlet *valves* and the other operating the exhaust valves. Gives improved engine efficiency due to improved fuel/air mixture and exhaust gas flow in the *combustion chambers*.

Two stroke (cycle) – A common term used to describe the operation of an engine where each downward *piston* stroke is a power stroke. The fuel/air mixture is directed to the crankcase where it's compressed by the descending piston and pumped into the *combustion chamber*. As the piston rises, the mixture is compressed and ignited, which forces the piston down. The burnt gases flow from the exhaust port, but the piston is now compressing another fuel/air mixture charge in the crankcase which repeats the cycle.

U

Understeer – A tendency for a car to go straight on when turned into a corner.

Universal joint – A joint that can swivel in any direction whilst at the same time transmitting *torque*. This type of joint is commonly used in *propeller shafts* and some *driveshafts*, but is not suitable for some applications because the input and output shaft speeds are not the same at all positions of angular rotation. The type in common use is known as a Hardy-Spicer, Hooke's or Cardan joint.

Unleaded petrol – Has no lead added during manufacture, but still has the natural lead content of crude oil. Generally available in the

CAR JARGON

UK, most modern cars can use this type of petrol, but seek advice first, as engine adjustments may be required. Engine damage can occur if unleaded petrol is used incorrectly. Not to be confused with *lead-free* petrol which is not currently available in the UK.
Unsprung weight – The part of the car which is not supported by the springs.

V

Vacuum advance – System of ignition *advance and retard* used in some *distributors* where the vacuum in the engine inlet *manifold* is used to act on a *diaphragm* which alters the *ignition timing* as the vacuum changes due to the throttle position.
Valve – A device which opens or closes to allow or stop gas or fluid flow.
Valve gear – A general term used for the components which are acted on by the *camshaft* in order to operate the *valves*.
16-valve – Term used to describe a four *cylinder* engine with four *valves* per cylinder (usually two inlet valves and two exhaust valves). Gives improved engine efficiency due to improved fuel/air mixture and exhaust gas flow in the *combustion chambers*.
Vee engine – An engine design in which the *cylinders* are set in two banks forming a 'V' when viewed from one end. A V8 for example consists of two rows of four cylinders each.
Venturi – A streamlined restriction in the *carburettor* throttle bore which causes a low pressure to occur; this sucks fuel into the air stream to form a vapour suitable for combustion.
Viscosity – A term used to describe the resistance of a fluid to flow. When associated with lubricating oil, it's given an *SAE* number, 10 being a very light oil and 140 being a very heavy oil.
Voltage regulator – A device which regulates the *alternator* output to a predetermined level. On most alternators the voltage regulator is an integral part of the alternator, and regulates the charging current as well as the voltage.

W

Wankel engine – A rotary engine which has a triangular shaped rotor which performs the function of the *pistons* in a conventional engine, and rotates in a housing shaped approximately like a broad-waisted figure of eight. Very few cars use this type of engine.
Wheel balancing – Adding small weights to the rim of a wheel so that there are no out-of-balance forces when the wheel rotates.
Wishbone – An 'A'-shaped *suspension* component, pivoted at the base of the 'A' and carrying a wheel at the apex. Normally mounted close to the horizontal.
Worm and nut steering – A steering system where the lower end of the steering column has a coarse screw thread on which a nut runs. The nut is attached to a spindle which carries the drop arm which, in turn, moves the steering linkage.

LOCAL RADIO FREQUENCIES

137

LOCAL RADIO

A comprehensive network of local radio stations now exists throughout the UK.

Most of these radio stations provide regularly updated reports on traffic flow and road conditions, which can be of great help to drivers. Listening to traffic reports will help you to avoid the inevitable 'jams' which occur during every day driving.

Some car radio/cassette players are now equipped with a 'Radio Data System' (RDS) which will automatically tune into special traffic information signals, interrupting normal radio or tape listening when bulletins are broadcast. An RDS-equipped radio may prove to be a worthwhile investment if you travel by car regularly, particularly when driving on business trips.

The following table provides details of all the local radio frequencies throughout the UK, region-by-region. Where stations broadcast on FM/VHF and AM/MW, it's suggested that the FM/VHF frequency is used wherever possible, as this will usually give better reception.

FORD FIESTA

LOCAL RADIO FREQUENCIES

AREA	FM/VHF	AM/MW
AVON		
Bath		
BBC Bristol	104.6	1548
GWR FM	103.0	–
Bristol		
BBC Bristol	94.9	1548
Brunel Classic Gold	–	1260
Galaxy Radio	97.2	–
GWR FM	96.3	–
BEDFORDSHIRE		
Bedford		
BBC Bedfordshire	95.5	1161
Chiltern Radio	96.9	–
SuperGold	–	792
Luton		
BBC Bedfordshire	103.8	630
Chiltern Radio	97.6	–
SuperGold	–	828
BERKSHIRE		
Reading		
210 Classic Gold Radio	–	1431
210 FM	97.0	–
BBC Berkshire	104.4	–
Wokingham		
BBC Berkshire	104.1	–
BIRMINGHAM		
Birmingham		
BBC WM	95.6	1458
BRMB FM	96.4	–
Buzz FM	102.4	–
Xtra AM	–	1152
BORDERS		
Eyemouth		
Radio Borders	103.4	–
Peebles		
Radio Borders	103.1	–
Selkirk		
Radio Borders	96.8	–
BUCKINGHAMSHIRE		
Milton Keynes		
BBC Bedfordshire	104.5	630
Horizon Radio	103.3	–

AREA	FM/VHF	AM/MW
CAMBRIDGESHIRE		
Cambridge		
BBC Cambridge	96.0	1026
CN FM	103.0	–
Peterborough		
BBC Cambridge	95.7	1447
Hereward Radio	102.7	1332
CHANNEL ISLANDS		
Guernsey		
BBC Guernsey	93.2	1116
Jersey		
BBC Jersey	88.8	1026
CHESHIRE		
Echo 96	96.4	–
Chester		
BBC Merseyside	95.8	1485
Congleton		
Signal Cheshire	104.9	–
Macclesfield		
BBC Stoke	94.6	1503
Warrington		
BBC Merseyside	95.8	1485
CLEVELAND		
Middlesbrough		
BBC Cleveland	95.0	1548
TFM	96.6	–
GNR	–	1170
CLWYD		
Prestatyn		
BBC Cymru/Wales	94.2	882
Rhyl		
BBC Cymru/Wales	94.2	882
Wrexham		
BBC Cymru/Wales	93.3	657
MFM 1034	97.1/103.4	–
Marcher Gold	–	1260
CORNWALL		
Isles of Scilly		
BBC Cornwall	96.0	–
Liskeard		
BBC Cornwall	95.2	657
Redruth		
BBC Cornwall	103.9	630
CUMBRIA		
Barrow-in-Furness		
BBC Furness	96.1	837
Kendal		
BBC Cumbria	95.2	–
Windermere		
BBC Cumbria	104.2	–
Workington		
BBC Cumbria	95.6	1458

AREA	FM/VHF	AM/MW
DERBYSHIRE		
Chesterfield		
BBC Sheffield	94.7	1035
Derby		
BBC Derby	94.2	1116
Trent FM	102.8	–
Gem AM	–	945
Matlock		
BBC Derby	95.3	–
DEVON		
Barnstaple		
BBC Devon	94.8	801
Exeter		
BBC Devon	95.8	990
DevonAir	97.0	666/954
South West	103.0	–
Okehampton		
BBC Devon	96.0	801
Plymouth Area		
BBC Devon	103.4	855
Plymouth Sound	97.0	1152
Tavistock		
Radio in Tavistock	96.6	–
Torbay		
BBC Devon	103.4	1458
DevonAir	96.4	666/954
DORSET		
Poole		
BBC Solent	96.1	1359
Bournemouth		
BBC Solent	96.1	1359
2CR Classic Gold		828
2CR FM	102.3	–
Weymouth		
BBC Solent	96.1	–
DUMFRIES & GALLOWAY		
Dumfries		
BBC Scotland/Solway	94.7	585
SW Sound	97.2	–
Stranraer		
BBC Scotland	94.7	–
DURHAM		
Darlington		
BBC Newcastle	95.4	–
Durham		
BBC Newcastle	95.4	1458
DYFED		
Aberystwyth		
BBC Cymru/Wales	93.1	882

LOCAL RADIO FREQUENCIES   139

AREA	FM/VHF	AM/MW
EAST SUSSEX		
Brighton		
BBC Sussex	95.3	1485
South Coast Radio	–	1323
Southern Sound Classic Hits	103.5	–
Eastbourne		
BBC Sussex	104.5	1161
Southern Sound Classic Hits	102.4	–
Hastings		
Southern Sound Classic Hits	97.5	–
Newhaven		
Southern Sound Classic Hits	96.9	–
ESSEX		
Basildon		
BBC Essex	95.3	765
Chelmsford		
BBC Essex	103.5	765
Breeze AM		1431/1359
Essex Radio	102.6	–
Colchester		
BBC Essex	103.5	729
Mellow 1557	–	1557
Harlow		
BBC Essex	–	765
Manningtree		
BBC Suffolk	103.9	–
Southend-on-Sea		
BBC Essex	95.3	1530
Breeze AM		1431/1359
Essex Radio	96.3	–
FIFE		
Kirkcaldy		
BBC Scotland	94.3	810
GLOUCESTERSHIRE		
Gloucester		
BBC Gloucester	104.7	603
Severn Sound	102.4	
Three Counties Radio		774
Stroud		
BBC Gloucester	95.0	603
Severn Sound	103.0	–
GRAMPIAN		
Aberdeen		
BBC Scotland/Aberdeen	93.1	990
North Sound	96.9	1035
Stirling		
Central FM	96.7	–

AREA	FM/VHF	AM/MW
GWENT		
Newport		
Red Dragon Radio	97.4	–
Touch AM		1305
GWYNEDD		
Anglesey		
BBC Cymru/Wales	94.2	882
HAMPSHIRE		
Andover		
210 FM	102.9	–
Basingstoke		
210 FM	102.9	–
BBC Berkshire	104.1	–
Isle of Wight		
BBC Solent	96.1	999/1359
Isle of Wight Radio	–	1242
Portsmouth		
BBC Solent	96.1	999
Oceansound	97.5	–
South Coast Radio	–	1170
Southampton		
BBC Solent	96.1	999
Oceansound	97.5	–
Power FM	103.2	–
South Coast Radio	–	1557
Winchester		
Oceansound	96.7	–
Power FM	103.2	–
HEREFORDSHIRE		
Hereford		
BBC Here/Worc	94.7	819
Radio Wyvern	97.6	954
HERTFORDSHIRE		
Watford		
BBC GLR	94.9	1458
St Albans		
BBC Bedfordshire	103.8	630
Stevenage		
BBC Bedfordshire	103.8	630
HIGHLAND		
Inverness		
BBC Scotland	94.0	810
Moray Firth Radio	97.4	1107
Fort William		
BBC Scotland	93.7	–
Oban		
BBC Scotland	93.3	–
Thurso		
BBC Scotland	94.5	–

AREA	FM/VHF	AM/MW
HUMBERSIDE		
Kingston upon Hull Area		
BBC Humberside	95.9	1485
Classic Gold	–	1161
Viking FM	96.9	–
ISLE OF MAN		
Douglas		
Manx Radio	97.2	1368
Snaefell		
Manx Radio	89.0	–
KENT		
Ashford		
Invicta FM	96.1	–
Canterbury		
BBC Kent	104.2	774
Invicta FM	102.8	–
Dover		
BBC Kent	104.2	774
Invicta FM	97.0	–
East Kent		
Coast Classics	–	603
Folkestone		
BBC Kent	104.2	774
Maidstone & Medway		
BBC Kent	96.7	1035
Coast Classics		1242
Invicta FM	103.1	–
Royal Tunbridge Wells		
BBC Kent	96.7	1602
Thanet		
Invicta FM	95.9	–
LANCASHIRE		
Burnley Area		
BBC Lancashire	95.5	855
Blackpool		
BBC Lancashire	103.9	855
Red Rose Gold	–	999
Red Rose Rock FM	97.4	–
Lancaster		
BBC Lancashire	104.5	1557
Preston		
BBC Lancashire	103.9	855
Red Rose Gold	–	999
Red Rose Rock FM	97.4	–
LEICESTERSHIRE		
Leicester		
BBC Leicester	104.9	837
Gem AM	–	1260
Sound FM	103.2	–

FORD FIESTA

LOCAL RADIO FREQUENCIES

AREA	FM/VHF	AM/MW
LINCOLNSHIRE		
Lincoln Area		
BBC Lincoln	94.9	1368
Lincs FM	102.2	–
LONDON		
Brixton		
Choice FM	96.9	–
Ealing		
Sunrise Radio	–	1413
Greater London		
BBC GLR	94.9	1458
Capital FM	95.8	–
Capital Gold	–	1548
Jazz FM	102.2	–
Kiss FM	100.0	–
LBC Newstalk	97.3	–
Melody Radio	104.9	–
Spectrum Int. Radio	–	558/990
Haringey		
London Greek Radio	103.3	–
WNK	103.3	–
Thamesmead		
London Talkback Radio	–	1152
RTM	103.8	–
Southall		
Sunrise Radio	–	1413
LOTHIAN		
Bathgate		
Radio Forth	97.3	–
Dunfermline		
BBC Scotland	94.3	810
Edinburgh		
BBC Scotland	94.3	810
Max AM	–	1548
Radio Forth	97.3/97.6	–
MANCHESTER		
Manchester		
BBC GMR	95.1	1458
KFM	104.9	–
Piccadilly Gold	–	1152
Piccadilly Key 103	103.0	–
Sunset Radio	102.0	–
MERSEYSIDE		
Liverpool		
BBC Mersey	95.8	1485
City Talk	–	1548
Radio City	96.7	–
MID GLAMORGAN		
Aberdare		
BBC Cymru/Wales	93.6	882
Merthyr Tydfil		
BBC Cymru/Wales	96.8	882

AREA	FM/VHF	AM/MW
MIDDLESEX		
Hounslow		
Sunrise Radio	–	1413
NORFOLK		
King's Lynn		
BBC Norfolk	104.4	873
Norwich Area		
BBC Norfolk	95.1	855
Broadland	102.4	1152
NORTHAMPTONSHIRE		
Kettering		
KCBC	–	1530
Northampton		
BBC Northampton	104.2	1107
Northants	96.6	–
SuperGold	–	1557
NORTHUMBERLAND		
Berwick-upon-Tweed		
BBC Newcastle	96.0	–
Radio Borders	97.5	–
NOTTINGHAM		
Nottingham		
BBC Nottingham	103.8	1521
Trent FM	96.2/96.5	–
Gem AM	–	999
ORKNEY		
Kirkwall		
BBC Scotland	93.7	–
OXFORDSHIRE		
Banbury		
Fox FM	97.4	–
Oxford		
BBC Oxford	95.2	1485
Fox FM	102.6	–
POWYS		
Welshpool		
BBC Cymru/Wales	94.0	882
SHETLAND		
Lerwick		
BBC Scotland	92.7	–
SIBC	96.2	–
SHROPSHIRE		
Ludlow		
BBC Shropshire	95.0	1584
RFM	97.6	–
Shrewsbury		
BBC Shropshire	96.0	–
Beacon Radio	97.2/103.1	
WABC	–	1017
Telford		
BBC Shropshire	96.0	756
Beacon Radio	97.2/103.1	
WABC	–	1017

AREA	FM/VHF	AM/MW
SOMERSET		
Mendip Area		
Orchard FM	102.6	–
Wells		
BBC Bristol	95.5	1548
SOUTH GLAMORGAN		
Cardiff		
BBC Cymru/Wales	96.8	882
Red Dragon Radio	103.2	–
Touch AM		1359
STAFFORDSHIRE		
Stafford		
Echo 96	96.9	–
Stoke on Trent		
BBC Stoke	94.6	1503
Signal Radio	102.6	1170
STRATHCLYDE		
Ayr		
West Sound	96.7	1035
Girvan		
West Sound	97.5	–
Glasgow		
BBC Scotland	94.3	810
Clyde 1	102.5	–
Clyde 2	–	1152
Radio Clyde FM	97.3	
Greenock		
BBC Scotland	94.3	810
Kilmarnock		
BBC Scotland	93.9	810
SUFFOLK		
Bury St. Edmunds		
Saxon Radio	96.4	1251
Great Barton		
BBC Suffolk	104.6	–
Ipswich		
BBC Suffolk	103.9	–
Radio Orwell	97.1	1170

FORD FIESTA

LOCAL RADIO FREQUENCIES — 141

AREA	FM/VHF	AM/MW
SURREY		
Guildford		
BBC Surrey	104.6	–
Delta Radio	97.1	–
First Gold Radio	–	1476
Premier Radio	96.4	–
Reigate		
Radio Mercury	102.7	1521
TAYSIDE		
Dundee		
Radio Tay	102.8	1161
BBC Scotland	92.7	810
Perth		
Radio Tay	96.4	1584
TYNE & WEAR		
Fenham		
BBC Newcastle	104.4	–
Metro FM	103.0	–
Newcastle Area		
BBC Newcastle	95.4	1458
GNR	–	1152
Metro FM	97.1	–
Sunderland		
Wear FM	103.4	–
ULSTER		
Belfast		
BBC Ulster	94.5	1341
Classic Trax BCR	96.7	–
Cool FM	97.4	–
Downtown Radio	102.6	–
Enniskillen		
BBC Ulster	93.8	873
Downtown Radio	96.6	–
Limavady		
Downtown Radio	96.4	–
Londonderry		
BBC Foyle	93.1	792
Downtown Radio	102.4	–
WARWICKSHIRE		
Leamington Spa		
Mercia FM	102.9	–

AREA	FM/VHF	AM/MW
WEST GLAMORGAN		
Swansea		
BBC Cymru/Wales	93.9	882
Swansea Sound	96.4	1170
WEST MIDLANDS		
Coventry		
BBC CWR	94.8	–
Mercia FM	97.0	–
Radio Harmony	102.6	–
Xtra AM	–	1359
Sutton Coldfield		
BBC Derby	104.5	–
Beacon Radio	97.2/103.1	–
Wolverhampton		
BBC WM	95.6	828
Beacon Radio	97.2/103.1	–
WABC	–	990
WEST SUSSEX		
Crawley		
Radio Mercury	102.7	1521
Horsham		
BBC Sussex	95.1	1368
Worthing		
BBC Sussex	95.3	1485
WESTERN ISLES		
Stornoway (Lewis)		
BBC Scotland	94.2	–
WILTSHIRE		
Chippenham		
BBC Wiltshire Sound	104.3	–
Marlborough		
GWR FM	96.5	–
Salisbury		
BBC Wiltshire Sound	103.5	–
Swindon		
BBC Wiltshire Sound	103.6	1368
Brunel Classic Gold	–	1161
GWR FM	97.2	–
West Wilts		
Brunel Classic Gold	–	936
GWR FM	102.2	–
WORCESTERSHIRE		
Worcester		
BBC Here/Worc	104.0	738
Radio Wyvern	102.8	1530

AREA	FM/VHF	AM/MW
YORKSHIRE NORTH		
Northallerton		
BBC York	104.3	–
Scarborough		
BBC York	95.5	1260
Whitby		
BBC Cleveland	95.8	–
York		
BBC York	103.7	666
YORKSHIRE SOUTH		
Barnsley		
Classic Gold	–	1305
Hallam	102.9	–
Doncaster		
BBC Sheffield	104.1	1035
Classic Gold	–	990
Hallam	103.4	–
Rotherham		
BBC Sheffield	88.6	1035
Hallam	96.1	–
Sheffield		
BBC Sheffield	88.6	1035
Classic Gold	–	1548
Hallam	97.4	–
YORKSHIRE WEST		
Bradford		
Classic Gold	–	1278
Pennine	97.5	–
Sunrise FM	103.2	–
Halifax		
Classic Gold	–	1530
Pennine	102.5	–
Huddersfield		
Classic Gold	–	1530
Pennine	102.5	–
Leeds		
Aire FM	96.3	–
BBC Leeds	92.4	774
Magic 828	–	828

FORD FIESTA

CONVERSION FACTORS

Length (distance)

Inches (in)	x 25.4	= Millimetres (mm)	x 0.0394	= Inches (in)
Feet (ft)	x 0.305	= Metres (m)	x 3.281	= Feet (ft)
Miles	x 1.609	= Kilometres (km)	x 0.621	= Miles

Volume (capacity)

Cubic inches (cu in; in³)	x 16.387	= Cubic centimetres (cc; cm³)	x 0.061	= Cubic inches (cu in; in³)
Pints (pt)	x 0.568	= Litres (l)	x 1.76	= Pints (pt)
Quarts (qt)	x 1.137	= Litres (l)	x 0.88	= Quarts (qt)
Gallons (gal)	x 4.546	= Litres (l)	x 0.22	= Gallons (gal)

Mass (weight)

Ounces (oz)	x 28.35	= Grams (g)	x 0.035	= Ounces (oz)
Pounds (lb)	x 0.454	= Kilograms (kg)	x 2.205	= Pounds (lb)

Force

Pounds-force (lbf; lb)	x 4.448	= Newtons (N)	x 0.225	= Pounds-force (lbf; lb)
Newtons (N)	x 0.1	= Kilograms-force (kgf; kg)	x 9.81	= Newtons (N)

Pressure

Pounds-force per square inch (psi; lbf/in²; lb/in²)	x 0.070	= Kilograms-force per square centimetre (kgf/cm²; kg/cm²)	x 14.223	= Pounds-force per square inch (psi; lbf/in²; lb/in²)
Pounds-force per square inch (psi; lbf/in²; lb/in²)	x 0.068	= Atmospheres (atm)	x 14.696	= Pounds-force per square inch (psi; lbf/in²; lb/in²)
Pounds-force per square inch (psi; lbf/in²; lb/in²)	x 0.069	= Bars	x 14.5	= Pounds-force per square inch (psi; lbf/in²; lb/in²)

CONVERSION FACTORS

Torque (moment of force)

Pounds-force feet (lbf ft; lb ft)	x 0.138	= Kilograms-force metres (kgf m; kg m)	x 7.233	= Pounds-force feet (lbf ft; lb ft)
Pounds-force feet (lbf ft; lb ft)	x 1.356	= Newton metres (Nm)	x 0.738	= Pounds-force feet (lbf ft; lb ft)
Newton metres (Nm)	x 0.102	= Kilograms-force metres (kgf m; kg m)	x 9.804	= Newton metres (Nm)

Power

Horsepower (hp)	x 0.745	= Kilowatts (kW)	x 1.3	= Horsepower (hp)

Velocity (speed)

Miles per hour (miles/hr; mph)	x 1.609	= Kilometres per hour (km/h; kph)	x 0.621	= Miles per hour (miles/hr; mph)

Fuel consumption*

Miles per gallon (mpg)	x 0.354	= Kilometres per litre (km/l)	x 2.825	= Miles per gallon (mpg)

Temperature

Degrees Fahrenheit = (°C x 1.8) + 32

Degrees Celsius (Degrees Centigrade; °C) = (°F − 32) x 0.56

* Note: It is common practice to convert from miles per gallon (mpg) to litres/100 kilometres (l/100 km), where mpg x l/100 km = 282

144 DISTANCE TABLES

	LONDON	Aberdeen	Aberystwyth	Birmingham	Bournemouth	Brighton	Bristol	Cambridge	Cardiff	Carlisle	Chester	Derby	Dover	Edinburgh	Exeter	Fishguard	Fort William	Glasgow	Harwich	Holyhead
LONDON		806	341	179	169	89	183	84	249	484	298	198	117	608	275	422	818	634	114	420
Aberdeen	501		724	658	903	895	789	742	792	349	571	634	924	198	911	795	254	241	848	692
Aberystwyth	212	450		193	328	425	206	354	177	375	153	222	459	526	325	90	708	525	460	182
Birmingham	111	409	120		237	267	132	161	163	309	117	63	296	460	259	283	642	459	267	246
Bournemouth	105	561	204	147		148	124	251	195	554	351	306	278	705	132	375	887	703	283	463
Brighton	55	556	264	166	92		217	180	288	573	386	286	130	697	275	468	906	723	203	509
Bristol	114	490	128	82	77	135		233	71	439	237	195	314	591	121	251	722	589	314	340
Cambridge	52	461	220	100	156	112	145		290	417	269	155	204	544	351	436	750	566	106	391
Cardiff	155	492	110	101	121	179	44	180		443	237	225	385	587	192	180	776	608	359	341
Carlisle	301	217	233	192	344	356	273	259	275		222	304	602	151	571	444	333	150	523	341
Chester	185	355	95	73	218	240	147	167	147	138		114	415	373	357	243	555	372	375	137
Derby	123	394	138	39	190	178	121	96	140	189	71		315	436	315	312	637	454	261	251
Dover	73	574	285	184	173	81	195	127	239	374	258	196		726	391	539	935	752	203	538
Edinburgh	378	123	327	286	438	433	367	338	365	94	232	271	451		723	616	233	72	650	489
Exeter	171	566	202	161	82	171	75	218	119	355	222	196	243	449		372	904	721	389	460
Fishguard	262	494	56	176	233	291	156	271	112	276	151	194	335	383	231		777	594	533	272
Fort William	508	158	440	399	551	563	480	466	482	207	345	396	581	145	562	483		183	856	671
Glasgow	394	150	326	285	437	449	366	352	378	93	231	282	467	45	448	369	114		673	488
Harwich	71	527	286	166	176	126	195	66	223	325	233	162	126	404	242	331	532	418		497
Holyhead	261	430	113	153	288	316	211	243	212	212	85	156	334	304	286	169	417	303	309	
Hull	168	348	229	143	255	223	226	124	244	155	132	98	251	225	301	285	362	250	181	217
Inverness	537	105	482	445	594	590	528	497	524	253	387	430	601	159	603	529	66	176	563	463
Leeds	190	322	173	111	261	251	206	148	212	115	78	74	269	199	268	229	322	210	214	163
Leicester	98	417	151	39	166	153	112	68	139	208	94	28	171	294	187	207	415	301	134	183
Lincoln	136	382	186	85	217	191	163	86	190	180	123	51	213	259	238	245	387	273	152	208
Liverpool	205	334	104	94	235	260	164	184	164	116	17	88	278	210	239	160	323	211	250	94
Manchester	192	332	133	81	228	247	167	154	172	115	38	58	265	209	242	189	322	210	220	123
Newcastle	271	230	270	209	348	326	295	231	310	58	175	164	344	107	370	326	257	143	297	260
Norwich	107	487	267	155	212	162	221	60	237	287	210	139	169	364	277	331	494	380	63	299
Nottingham	123	388	154	50	189	178	132	84	152	185	87	16	196	265	207	210	392	278	150	172
Oxford	56	464	157	62	92	99	66	79	107	256	128	98	129	341	139	207	461	340	134	202
Penzance	282	683	313	268	193	269	186	329	230	466	333	307	354	560	111	342	673	559	353	397
Plymouth	213	614	244	199	124	213	117	260	161	397	264	238	285	491	42	273	604	490	284	328
Preston	216	302	144	105	256	271	185	198	190	85	49	100	289	179	266	200	292	180	264	125
Sheffield	162	358	158	75	227	217	163	124	176	145	78	36	235	235	248	214	352	240	190	163
Southampton	79	530	202	128	32	60	75	136	119	322	194	164	149	416	111	231	520	415	150	283
Stranraer	410	235	342	301	453	465	382	368	384	109	247	298	483	133	464	385	199	85	434	321
Swansea	196	506	76	126	162	220	85	216	41	289	151	165	280	383	160	71	496	382	264	189
Worcester	114	428	98	26	131	168	60	116	75	211	88	65	187	305	135	155	418	304	182	152
York	196	309	197	134	273	251	215	157	242	116	102	89	269	186	290	253	323	211	223	187

MILES

FORD FIESTA

DISTANCE TABLES 145

KILOMETRES

Hull	Inverness	Leeds	Leicester	Lincoln	Liverpool	Manchester	Newcastle	Norwich	Nottingham	Oxford	Penzance	Plymouth	Preston	Sheffield	Southampton	Stranraer	Swansea	Worcester	York	
270	864	306	158	219	330	309	436	172	198	90	454	343	348	261	127	660	315	183	315	**LONDON**
560	169	518	671	615	538	534	370	784	624	747	1099	988	486	576	853	378	814	689	497	Aberdeen
369	776	278	243	304	167	214	435	430	248	253	504	393	232	254	325	550	122	158	317	Aberystwyth
230	716	179	63	137	151	130	336	249	80	100	431	320	169	121	206	484	203	42	216	Birmingham
410	956	420	267	349	378	367	560	341	304	148	311	200	412	365	51	729	261	211	439	Bournemouth
359	950	404	246	307	418	398	525	261	286	159	433	343	436	349	97	748	354	270	404	Brighton
364	850	332	180	262	264	269	475	356	212	106	299	188	298	262	121	615	137	97	346	Bristol
200	800	238	109	138	296	248	372	97	135	127	529	418	319	200	219	592	348	187	253	Cambridge
393	843	341	224	306	264	277	499	381	245	172	370	259	306	283	192	618	66	121	389	Cardiff
249	407	185	335	290	187	185	93	462	298	412	750	639	137	233	518	175	465	340	187	Carlisle
212	623	126	151	198	27	61	282	338	140	206	536	425	79	126	312	398	243	142	164	Chester
158	692	119	45	82	142	93	264	224	26	158	494	383	161	58	264	480	266	105	143	Derby
404	967	433	275	343	447	426	554	272	315	208	570	459	465	378	240	777	451	301	433	Dover
362	256	320	473	417	338	336	172	586	426	549	901	790	288	378	669	214	616	491	299	Edinburgh
484	970	431	301	383	385	389	595	446	333	224	179	68	428	399	179	747	257	217	467	Exeter
459	851	369	333	394	257	304	525	533	338	333	550	439	322	344	372	620	114	249	407	Fishguard
583	106	518	668	623	520	518	414	795	631	742	1083	972	470	566	837	320	798	673	520	Fort William
402	283	338	484	439	340	338	230	612	447	547	900	789	290	386	668	137	615	489	340	Glasgow
291	906	344	216	245	402	354	478	101	241	216	568	457	425	306	241	698	429	293	359	Harwich
349	745	262	295	335	151	198	418	481	277	325	639	528	201	262	455	517	304	245	301	Holyhead
	615	90	145	61	196	151	190	240	148	262	663	552	180	103	412	425	433	272	63	Hull
382		579	729	673	605	604	428	842	682	816	1159	1049	555	634	925	417	872	747	552	Inverness
56	358		158	108	119	64	148	277	116	272	613	502	90	58	378	360	369	220	39	Leeds
90	453	98		82	187	138	301	187	42	119	480	369	206	105	225	510	282	109	174	Leicester
38	418	67	51		192	137	245	169	56	201	562	451	185	72	307	465	340	192	124	Lincoln
122	376	74	116	119		55	253	360	167	251	563	452	50	119	357	362	270	169	158	Liverpool
94	375	40	86	85	34		212	306	119	230	568	457	48	64	325	360	295	164	103	Manchester
118	266	92	187	152	157	132		414	254	412	774	663	203	206	518	262	517	378	127	Newcastle
149	523	174	116	105	224	190	257		198	224	626	515	354	241	299	637	439	283	293	Norwich
92	424	72	26	35	104	74	158	123		156	512	401	167	64	262	473	283	122	134	Nottingham
163	507	169	74	125	156	143	256	139	97		402	291	272	217	106	587	230	95	291	Oxford
412	720	381	298	349	350	353	481	389	318	250		126	597	560	357	925	436	396	645	Penzance
343	652	312	229	280	281	284	412	320	249	181	78		486	447	246	814	325	285	534	Plymouth
112	345	56	128	115	31	30	126	220	104	169	371	302		113	375	312	322	203	126	Preston
64	394	36	65	45	74	40	128	150	40	135	348	278	70		323	490	336	163	90	Sheffield
256	575	235	140	191	222	202	322	186	163	66	222	153	233	201		679	257	193	398	Southampton
264	259	224	317	289	225	224	163	396	294	365	575	506	194	254	422		641	515	362	Stranraer
269	542	229	175	211	168	183	321	273	176	143	271	202	200	209	160	398		161	418	Swansea
169	464	137	68	119	105	102	235	176	76	59	246	177	126	101	120	320	100		257	Worcester
39	343	24	108	77	98	64	79	182	83	181	401	332	78	56	247	225	260	160		York

FORD FIESTA

INDEX

A

About this Handbook 5
Accidents 29
   Dealing with 29
   First aid 30
   Recording details of 32
   Requirements of the law 32
Air cleaner element, renewal 97
Air cleaner intake spout 102, 103
Alcohol and driving 48
Alternator drivebelt, checking 95, 102
Asbestos, precautions concerning 79
Automatic transmission 23, 76, 96
   Fluid level 92
Axle stands 80, 81

B

Battery:
   Connections 94, 102
   Electrolyte level 70, 102
   Precautions concerning 79
   Starting when flat 41
Bodywork, checks and
maintenance 90, 102, 103, 105, 106
Bonnet release lever 25, 89
Boot light, bulb renewal 114
Brake fluid, low level warning light 21, 89
Brake fluid level 68
Brake light, bulb renewal 110
Braking system, checks 89, 92, 96, 119
Breakdowns 37
   Emergency kit 42
Broken windscreen, dealing with 36
Bulbs, renewal 107
   Boot light 114
   Courtesy light 114
   Direction indicator (front) 109
   Direction indicator repeater light 110
   Driving light 110
   Glovebox light 114
   Headlight 108
   Luggage compartment light 114
   Number plate light 113
   Rear foglight 112
   Rear lights 110, 111
   Reversing light 112
   Sidelight (front) 109
   Tail lights 110, 111
Bump starting 41
Buying and selling 9

C

Capacities, lubricants and fluids 62
Car crime prevention 59
Car jargon .... 137 to 141
Caravan, towing 46
Changing a wheel 38
Child safety 51
Choke 22
Cigarette lighter 26
Clock 19, 20
Contact breaker points:
   Gap adjustment 63, 88
   Renewal 90
Controls and equipment 15
Conversion factors 142
Coolant:
   Checking strength 101
   Level 67
   Renewal 98
Cooling system, flushing 99
Courtesy light, bulb renewal 114
Crankcase emission control filter (CVH engine).
.... 97
Crime prevention 59
CTX (automatic) transmission 23, 76, 92, 96
CVH engine identification 83

D

Dimensions and weights 13
Direction indicator (front), bulb renewal 109
Direction indicator repeater light, bulb renewal
.... 110
Direction indicator warning light 21
Distance tables 144
Distributor, check/clean/lubricate 85, 87
Door locks 28, 89
Drivebelt (alternator), checking 95
Driving:
   abroad 53
   in bad weather 44
   on a motorway 45, 46
   safety 43
Driving light, bulb renewal 110

FORD FIESTA

INDEX

E

Econolights 21
Economical driving 57
Electrical system, checks 94, 96
Electricity (mains) and electrical equipment,
precautions concerning 80
Emergencies 29
Engine compartment layouts 66, 82, 83
Engine oil and filter, renewal 84
Engine oil, disposal of 80
Environmental protection 57, 80
Equipment 15, 102

F

Fault finding 121
    Brakes 124
    Electrics 126
    Engine .... 121 to 123
    Gearbox, transmission and clutch 123
    Suspension and steering 125
Fiesta family 7
Filter (air), renewal 97
Filter (oil), renewal 84
Fire hazards, precautions concerning 79
Fire in vehicle, dealing with 36
First aid 30
Foglight switch (rear) 25
Front quarterlight 26
Front seat adjustment 27
Fuel gauge 19
Fuel system, checks and adjustments
.... 97, 102, 103
Fumes, precautions concerning 79
Fuses, renewal 114

G

Gearbox (manual) 22
    Oil level 91
Glossary .... 127 to 136
Glovebox light, bulb renewal 114

H

Handbrake, servicing .... 92 to 94
Handbrake warning light 21
Hazard flasher (warning) switch 25
Head restraints 27
Headlight, bulb renewal 108
Headlight main beam warning light 21
Heated rear window switch 25
Heating and ventilation controls 24
Hinges 89
Horn 23, 72
Hoses 84, 101
HT leads, check and clean 85

I

Identifying engine types 83
Ignition HT voltage, precautions concerning .. 80
Ignition switch/steering lock 22
Ignition system checks 85, 90
Ignition warning light 21
Instrument illumination control 25
Instrument panel layouts .... 15 to 18
Interior, cleaning 105

J

Jacking and vehicle support 38, 80
Jacking and wheel changing 38
Jargon .... 127 to 136
Jump lead starting 41

L

Leaks (fluid), checking for 72
Lights, checking operation 72, 102, 119
Lights switch (exterior lights) 23
Local radio frequencies .... 137 to 141
Locks (door) 28, 89
Loudspeaker balance control 25
Lubricants and fluids 61
Luggage compartment light, bulb renewal . 114

FORD FIESTA

INDEX

M

Maintenance schedule – see Service schedule
Mirror (interior rear view) 27
MOT, preparing for 119
   Brakes 119
   General 120
   Lights 119
   Seatbelts 120
   Steering and suspension 119
   Tyres and wheels 120
Motorway driving 45, 46
Multi-function switch 23

N

Number plate light, bulb renewal 113

O

OHV engine identification 83
Oil and filter, renewal 84
Oil filler cap (OHV engine) 96
Oil level 66
Oil pressure warning light 21
Oil, safe disposal of 80
Oil types (recommended) 61

P

Paintwork, scratches 90, 106
Parcel shelf 28
Pipes 84
Plugs (spark) 63, 86, 90
Points (contact breaker):
   Gap adjustment 63, 88
   Renewal 90
Punctures 38
Push starting 41

Q

Quarterlight 26

R

Radio frequencies .... 137 to 141
Rear foglight switch 25
Rear lights, bulb renewal 110, 112
Rear number plate light, bulb renewal 113
Rear parcel shelf 28
Rear seats 27
Rear view mirror (interior) 27
Recommended lubricants and fluids 61
Reducing the cost of motoring 57
Regular checks 65
Relays, renewal 116
Remote tailgate release switch 25
Rev counter 19
Reversing light ('Mark1' models), bulb renewal .
.... 112
Road testing 96
Routine checks 65
Routine maintenance 73

S

Safety (child) 51
Safety (driving) 43
Safety first! 78
Seat adjustment 27
Seat belts 27, 89, 120
Seats (child) 52
Security 59
Selling 11
Service schedule .... 74 to 77
   250 miles (400 km)/weekly 84
   6000 miles (10000 km)/6 months 84
   12000 miles (20000 km)/12 months 90
   24000 miles (40000 km)/1 year 96
   24000 miles (40000 km)/2 years 96
   36000 miles (60000 km) 98
   36000 miles (60000 km)/2 years 98
   36000 miles (60000 km)/3 years 101
   Seasonal servicing 101
Service tasks 83
Servicing 73
Sidelight (front), bulb renewal 109
Sidelight/tail light warning light 21
Skid control 48
Spare parts 82
Spark plug specifications 63
Spark plugs, renewal 86, 90
Speaker balance control 25
Speedometer 19

FORD FIESTA

INDEX 149

Starting a car with a flat battery 41
Steering, checks 119
Steering column switches 23
Steering lock 22
Sunroof 26
Suspension, checks 119
Switches:
    Hazard flasher (warning) 25
    Heated rear window 25
    Ignition 22
    Lights 23
    Multi-function 23
    Rear foglight 25
    Tailgate remote release 25
    Tailgate wash/wipe 23
    Windscreen wash/wipe 23, 24

T

Tachometer 19
Tail lights, bulb renewal 110
Tailgate lock 28
Tailgate release switch 25
Tailgate wash/wipe switch 23
Temperature gauge 19
Tools 104
Tow starting 41
Towing 13, 40
Trailer, towing 46
Troubleshooting – see Fault finding
Tyre pressures 62
Tyres, checking 68, 102, 120

U

Underbody, checking and cleaning 90, 103
Underbonnet layouts 66, 82, 83

V

Vandalism, dealing with 36
Ventilation controls 24

W

Warning lights 21
    Brake fluid, low level 21, 89
    Choke 22
    Direction indicator 21
    Econolights 21
    Handbrake 21
    Headlight main beam 21
    Ignition 21
    Oil pressure 21
    Sidelights/tail lights 21
Wash/wipe switch 23
Washer fluid level 70, 102
Washer pump switch 24
Weekly checks 65
Weights 13
Wheel bolts, checking 89
Wheel changing 38
Windscreen breakage, dealing with 36
Windscreen wash/wipe switch 23
Wipers and washers, checking 70, 71, 102
Women drivers, advice to 49

FORD FIESTA

150 SERVICING NOTES

Date	Action

FORD FIESTA

SERVICING NOTES

Date	Action

152 SERVICING NOTES

Date	Action

FORD FIESTA

SERVICING NOTES

Date	Action

FORD FIESTA

SERVICING NOTES

Date	Action

FORD FIESTA

SERVICING NOTES

Date	Action

SERVICING NOTES

Date	Action

FORD FIESTA

SERVICING NOTES

Date	Action

SERVICING NOTES

Date	Action

FORD FIESTA

SERVICING NOTES

Date	Action

FORD FIESTA

160 OWNERS WORKSHOP MANUALS

Fiesta Manuals

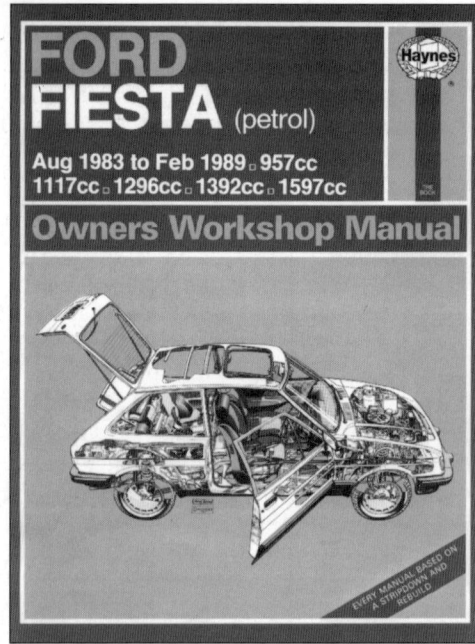

Like this handbook, the above manuals have been written specifically for your Fiesta. Most models are covered from 1976 to February 1989, the only exceptions being those fitted with Diesel engines (for which a separate manual on Ford Diesel engines is produced), CTX transmissions and the revised Fiesta range introduced in February 1989.

The manuals are written in the same easy-to-follow manner as the handbooks and enable you to tackle even major overhauls and repairs yourself. They will advise you on which jobs are suitable for the DIY mechanic and often suggest practical alternatives to the specialist tools that are so often recommended.

With the ever increasing cost of garage labour, think of the money you could save!

Haynes Manuals are available from most good motor accessory stores and bookshops, or direct from the publisher.

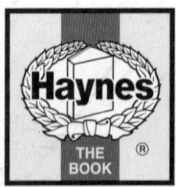 **HAYNES**

Haynes Publishing Group, Sparkford, Nr Yeovil, Somerset BA22 7JJ Telephone 0963 40635

FORD FIESTA